LEARNING MATHEMATICS THROUGH

LEARNING MATHEMATICS
THROUGH *DERIVE*»

**JOHN S. BERRY, EDWARD GRAHAM and
ANTONY J. P. WATKINS**
Centre for Teaching Mathematics,
University of Plymouth

Other books in the "Learning through Computer Algebra" Series
Published by Chartwell-Bratt. Series editor: John Berry.

"Learning Numerical Analysis through DERIVE", T. Etchells, J. Berry
"Learning Linear Algebra through DERIVE", B. Denton
"Learning Differential Equations through DERIVE" by B. Lowe, J. Berry
"Learning Modelling through DERIVE", S. Townend, D. Pountney.

British Library Cataloguing in Publication Data
A catalogue record for this book is available from the British Library

First published 1993 by Ellis Horwood Limited
Next published 1995 by Prentice Hall

© J. S. Berry, E. Graham, A. J. P. Watkins and Chartwell-Bratt 1996.

Chartwell-Bratt (Publishing and Training) Ltd
ISBN 0-86238-461-3

Printed in Sweden
Studentlitteratur, Lund
Art.nr. 6458
ISBN 91-44-00330-7

Printing/Year   1 2 3 4 5 6 7 8 9 10 | 2000 99 98 97 96

Contents

Series Preface

This series is designed to encourage the teaching and learning of various courses in mathematics using the symbolic algebra package DERIVE. Each text in the series will include the 'standard theory' appropriate to the subject, worked examples, exercises and *DERIVE Activities* so that students have a sound conceptual base on which to build. It is through the DERIVE activities that the learning of mathematics through investigations and discussion is encouraged. The aims of each DERIVE activity are to provide an investigation to introduce a new topic and to introduce the commands for using DERIVE as a manipulator in mathematical problem solving. Mathematics teaching and learning is changing as new technology becomes more readily available and these texts are written to support the new methods.

John Berry
Centre for Teaching Mathematics
University of Plymouth

Preface

The DERIVE User Manual begins:

> Making mathematics more exciting and enjoyable is the
> driving force behind the development of the **Derive** program.

The authors of DERIVE have succeeded! In using DERIVE for the past four years with our engineering and mathematics students we would agree that by removing the drudgery of long tedious manipulations and calculations our students are learning mathematics with more enthusiasm than before. The mathematics classroom has been transformed from only formal lectures and exercise classes to include interactive workshops with the students learning through investigation and discussion. This text is the result of our experiences.

The content of this text consists of the basic mathematics taught in nearly all first courses in mathematics for science and engineering. It is based on the experience of the three authors of teaching such material for many years. The text falls naturally into three sections: functions, calculus and algebra. Chapter 1 is essentially a review showing how to use DERIVE to explore mathematical concepts and to use it in modelling and mathematical problem solving. In chapters 2 and 3 we introduce the exponential, logarithmic and trigonometric functions. Chapter 4 is about sequences and series. Numerical methods are integrated into the text at the appropriate points through chapters 5 and 8. The bulk of the calculus takes up a third of the text in chapters 6, 7 and 9. For the engineer and scientist these chapters provide the mathematical tools for modelling many real situations. Chapters 10 and 11 look briefly at an introduction to complex numbers and matrices. (Another text in this DERIVE series explores the use of DERIVE in learning linear algebra.) In schools and colleges in the United Kingdom, this material is taught as part of GCE Advanced level and corresponding BTEC and GNVQ courses and in many Universities the material forms part of a foundation course in science and engineering.

The text contains the 'standard theory', worked examples, exercises and *DERIVE Activities*. It is through these DERIVE Activities that the learning of mathematics through investigation and discussion is encouraged. The aims of each DERIVE Activity are to provide an investigation to introduce a new topic (often before a more formal approach) and to introduce the commands for using DERIVE as a manipulator in mathematical problem solving. The DERIVE Activities could be used as a resource for workshop sessions with large groups, for self-learning a new topic, or reviewing a topic taught in the more traditional lecture/lesson.

What makes this text different from the many others on the market at this level is the use of the computer algebra package DERIVE. This powerful tool does for algebraic manipulation what the calculator does for arithmetic. However a word of warning! The authors believe that the introduction of DERIVE into the learning classroom *does not replace* the need to learn and understand algebraic manipulation. Our experience suggests that the *best* users of DERIVE are those students who are good at algebraic manipulation and understand the concepts behind algebra and calculus. The text attempts to blend the traditional skills with the power of DERIVE. To encourage good practice many of the exercises should be done 'by hand' first and then the results checked using DERIVE.

The appropriate use of technology is an important part of the learning process. The hand calculator is now accepted as a natural resource to have in the classroom and it is easy to carry around. Until recently the use of powerful software packages was restricted to the computer laboratory. This was often quoted as a disadvantage in the use of DERIVE; we were often told that engineering students would not have the access to DERIVE in their engineering classes! However technological change is rapid. DERIVE is now available on the HEWLETT-PACKARD HP 95LX palm top computer. Much of this text has been tried and tested using an HP95LX while on trains and planes - The power of DERIVE is now pocket size!

A natural question that we are often asked is "Why Derive?". The answer is quite easy! No other computer algebra software package then (and we believe since then) can match the combination of power and user friendliness than DERIVE. Within a two hour workshop session we have found students (both at school and in higher education) feeling 'at home' with the package using it confidently to explore mathematical concepts and to solve mathematical problems. Students are encouraged to do mathematics not watch it being done by the teacher. What else can the teacher of mathematics require?

We have assumed that the reader has had an introduction to DERIVE before coming to this text. The commands that you should be familiar with are **Author, Plot, Manage Substitute, Expand, Factor, Remove, Simplify, approX**. For complete newcomers to DERIVE, worksheets for an introductory workshop session are available from the authors.

Before beginning most DERIVE Activities you are recommended to set **Options: Precision** to **Mixed**.

We are extremely grateful to Sharon Ward for her skill and meticulous care in producing the camera ready copy for this text. We appreciate her patience and advice in the many changes to the page design and layout. We also thank Jenny Sharp for producing the artwork which has helped enhance the finished quality of the text.

We would welcome comments from teachers and students for improving this text. Since this is one of the first textbooks to fully integrate a computer algebra package into the learning of mathematics it is inevitable that the mix of DERIVE activities and standard theory could be improved. If you have constructive criticisms to help the writing of a second edition please write to the authors in Plymouth.

We hope that the teacher and learner will enjoy their mathematical activities using DERIVE.

John Berry
Ted Graham
Tony Watkins

Centre for Teaching Mathematics,
The University of Plymouth,
Drake Circus, Plymouth
Devon PL4 8AA, UK.
Telephone/FAX 0752 232772

1

Introducing functions

1.1   LINEAR LAWS

DERIVE ACTIVITY 1A

The aim of this Activity is to familiarise you with the algebra and plotting windows of DERIVE.

(A)   Load DERIVE. At the bottom of the screen you will see a line of commands, above them is the working area. Some information is displayed below the command line. In this activity you will examine some examples of linear functions. If you press P for **Plot**, then B for **Beside** and enter you will find the screen splits into two halves, known as windows. The left hand window will be used for algebraic work and the right hand window for plotting.

(B)   To plot points you need to define a pair of coordinates (x,y). The x values give the horizontal movement from the point at the centre of the screen where the two axes cross and the y value gives the vertical movement. Press M for **Move**. Change the x-value to 0 and then press Tab and change the y-value to 2. Now press enter, and the small cross moves to the point $(0,2)$. Move the cross to the points $(0,0)$, $(2,1)$, $(-1,1)$, $(-1,-2)$ and $(0,-3)$.

(C)   Press A for **Algebra** to move to the algebra window. The **Author** command is used to enter most expressions and data. Press A for **Author** and type [2,1]. Now press P for **Plot** to move to the plotting window and P for **Plot** again to actually plot the point.

Use this procedure to plot the points $(1,0)$, $(0,-1)$, $(-1,-2)$ and $(-2,-3)$.

What do you notice about these points? Return to the algebra window and **Author** the expression $y = x - 1$. Then move to the plotting window and **Plot** this expression. You should get a straight line that passes through all the points that you plotted before. Your DERIVE screen should look like Figure

1.1. When the graph produced is a straight line, as in this example, we say that x and y are related by a *linear law*.

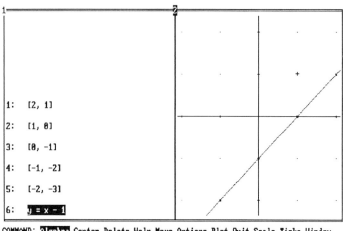

Figure 1.1

The equation $y = x - 1$ defines a relationship between the x and y coordinates. The y coordinate is obtained by subtracting 1 from the x coordinate. Clear the plotting window by typing D for **Delete**, A for **All** and enter.

(D) (i) Change the scale by pressing S for **Scale**, 2, Tab, 2, enter. Each mark of the axes now represents 2 units, note that this is displayed at the bottom of the screen.

(ii) Plot the points (4,5), (–3, –2) and (0,1). In the **Algebra** window **Author** $y = x + 1$ and then **Plot** this. Again the line should pass through the three points.

(iii) Repeat (ii) for:

(a) (2,5) (1,3) (–2,–3) and $y = 2x + 1$

(b) (0,0) (2,4) (–1,–2) and $y = 2x$

(c) (3,5) (1,1) (–2,–5) and $y = 2x - 1$

(iv)   You should have 3 straight lines all of which are parallel. What do the 3 equations have in common?

Where does each line cross the vertical axis? What do you notice if you compare these values with the equations?

(E)   (i)   Clear your plotting window using the **Delete All** commands. Now **Author** and **Plot** the lines with equations $y = x + 1$, $y = 2x + 1$, $y = 3x + 1$.

(ii)   What do the 3 lines have in common? What do the 3 equations have in common?

(iii) Which line is steepest? **Author** the expression **GRAD**$(3x+1,[x])$. Then press S for **Simplify** and enter. DERIVE gives you the gradient of the line $y = 3x + 1$. Repeat for $2x + 1$ and $x + 1$. What do you notice about the gradient and the equation?

In general the line with equation $y = mx + c$ has *gradient m* and crosses the vertical axis at c. The gradient tells you how steep the line is. The value c is often referred to as the *intercept*. Because the graph is a straight line we say that there is a linear law or relationship between x and y.

(F)   (i)   **Delete All** in the plotting window, then **Author** and **Plot** each of the straight lines defined by the equations below:

$$y = -x + 2 \qquad y = \tfrac{1}{2}x + 1$$

$$y = -2x + 4 \qquad y = \tfrac{1}{4}x - 1$$

$$y = -3x - 2 \qquad y = \tfrac{1}{5}x + 2$$

(ii)   Find the gradient of each line using the **GRAD** command. What does a negative gradient indicate? Also check that each line has the intercept that matches the equation.

1.2   THE EQUATION OF A STRAIGHT LINE

The equation of a straight line is given by $y = mx + c$ where m is the gradient of the line and c the intercept. The gradient gives the steepness of the line and is defined as

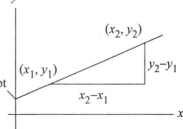

$$\text{gradient} = \frac{\text{change in } y}{\text{change in } x}.$$

If the coordinates of two points on the line are (x_1, y_1) and (x_2, y_2) then the gradient is given by

$$\frac{y_2 - y_1}{x_2 - x_1}.$$

Figure 1.2

The intercept is the point where the line cuts the vertical or y-axis.

Example 1A

A straight line passes through the points $(2,6)$ and $(1,4)$.

(i)   Find the gradient of the line.

(ii)   Find the equation of the line.

Solution

(i)   The gradient m is given by

$$m = \frac{y_2 - y_1}{x_2 - x_1} = \frac{6 - 4}{2 - 1} = \frac{2}{1} = 2.$$

(ii)   As the gradient of the line is 2 the equation must be of the form $y = 2x + c$. As the line passes through $(1,4)$ we can replace y by 4 and x by 1 to give

$$4 = 2 \times 1 + c$$
$$4 = 2 + c$$

so $c = 2$ and the equation is $y = 2x + 2$.

Example 1B

Find any three points that lie on the line $y = 5x - 3$ and then draw the line.

Solution

Taking $x = 1$ gives $y = 5 \times 1 - 3 = 2$ so $(1,2)$ is one point.
Taking $x = 2$ gives $y = 5 \times 2 - 3 = 7$ so $(2,7)$ is a second point.
Taking $x = 0$ gives $y = 5 \times 0 - 3 = -3$ so $(0,-3)$ is a third point.

Figure 1.3 shows these 3 points and a line drawn through them.

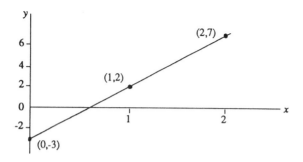

Figure 1.3

Exercise 1A

1.   For each equation given below find 3 points that lie on the line and then draw the line that passes through them.

(a) $y = 2x + 3$ (e) $y = -2x + 1$
(b) $y = x - 4$ (f) $y = 6x - 2$
(c) $y = 5x - 4$ (g) $y = -4x + 6$
(d) $y = \frac{1}{2}x + 1$ (h) $y = -2x - 4$

2.   Find the gradient and the equation of the line that passes through each pair of points given below.

(a) $(0,0)$ $(2,6)$ (e) $(3,0)$ $(6,1)$
(b) $(1,1)$ $(4,5)$ (f) $(5,4)$ $(0,5)$
(c) $(3,7)$ $(0,4)$ (g) $(1,4)$ $(3,-2)$
(d) $(0,6)$ $(2,2)$ (h) $(-1,6)$ $(5,-2)$

1.3   PROPORTIONALITY

If a linear equation is of the form

$$y = kx$$

then y and x are said to be *proportional* and in symbols this is often written as $y \propto x$ (y is proportional to x). The constant k is known as the *constant of proportionality*. If y is proportional to x then the graph of the equation $y = kx$ is a straight line with gradient k passing through the origin.

Example 1C

Table 1.1 below gives values for the tension in a spring and its extension.

Table 1.1

Tension T (N)	0	0.5	1.0	1.5	2
Extension e (cm)	0	4	8	12	16

Show that the tension is proportional to the extension and find the constant of proportionality.

Solution

Figure 1.4 shows a graph of tension against extension.

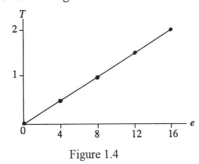

Figure 1.4

As we have a straight line passing through (0,0) then the two quantities must be proportional.

As $T \propto e$, then $T = ke$. When the tension is 16 the extension is 2, so $16 = k \times 2$, and $k = 8$.

Example 1D

The speed at which a stone hits the ground is proportional to the square root of the height from which it was released. A stone dropped from a height of 4 m hits the ground at 9 ms^{-1}. Find the constant of proportionality. Calculate the speed of the stone when it is dropped from a height of 8 m.

Solution

Using v for speed and h for height gives

$$v \propto \sqrt{h}$$

or

$$v = k\sqrt{h} \ .$$

Substituting in the known values for v and h gives

$$9 = k \times \sqrt{4} \ .$$

Solving for k we have

$$k = 4.5 \ .$$

Thus $v = 4.5\sqrt{h} \ .$

When $h = 8$ the speed is given by

$$v = 4.5\sqrt{8}$$
$$= 12.73 \ \mathrm{ms}^{-1} .$$

Exercise 1B

1. For each set of data given below, check to see if the two variables are proportional. If they are find the constant of proportionality.

(a)

P	0	1	3	5
Q	1	2	4	7

(c)

x	1	2	5	9
y	2	3	6	10

(b)

R	1	2.5	3	5
S	10	25	30	50

(d)

v	1.1	1.2	1.3	1.5	2
t	1.65	1.80	1.95	2.25	3.00

2. The tension in a spring is proportional to its extension. A spring stretches 8 cm under a tension of 5 N. Find the constant of proportionality. What tension would be present if the extension were 10 cm?

3. The petrol consumed by a car is proportional to the distance travelled. If 300 miles can be travelled using 60 litres of petrol, find the constant of proportionality. How far could the car travel on 100 litres of petrol?

4. The diameter of the stem of a plant is proportional to the square root of its growing time. A plant that has been growing for 36 days has a diameter of 0.9 cm. What will the diameter be after 100 days?

5. The friction force F acting on a sliding block is proportional to the normal reaction force R. Complete Table 1.2 and find the equation relating F and R.

Table 1.2

Normal Reaction R(N)	19.8	16.2	
Friction F(N)	6.1		4.7

6. The pressure P of a quantity of gas is proportional to its temperature, T and is proportional to its density, ρ. Write down an equation describing the law between P, T and ρ.

1.4   MODELLING WITH LINEAR LAWS

Often data collected from experiments lie close to a straight line which does not pass through the origin. For example Figure 1.5 shows a graph of the length ℓ of a heated copper bar (in centimetres) against the temperature T of the bar (in degrees celsius).

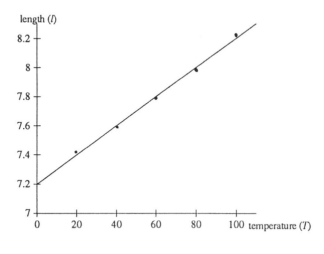

Figure 1.5

The data points are very close to a straight line called the 'line of best fit'. A simple rule for drawing this line is to make sure that there are as many points on one side of the line as the other.

The equation of the straight line is found by calculating its gradient and intercept

$$\ell = 0.01T + 7.2 \ .$$

We say that there is a *linear relationship* or a *linear model* between the variables ℓ and T. The most commonly occurring models in science and engineering are linear models.

The following example shows how we use DERIVE to find linear models between variables.

Example 1E

In an experiment to investigate the tension in an elastic string the tension in the string and length of the string were measured and the results shown in Table 1.3.

Table 1.3 Experimental data investigating tension

length of spring λ (in centimetres)	20.0	38.5	48.0	57.0	67.0
tension in spring T (in newtons)	10.0	17.5	21.5	25.5	29.0

Show that a linear law is a good model relating T and λ and find the equation of the 'line of best fit'.

Solution

Load DERIVE and open a plot window.

The first step is to plot the graph of the data. Reading in the data values in pairs, the graph suggests that λ and T are linearly related (see Figure 1.6 in which the scales have been changed to x:40 and y:15).

The DERIVE function **FIT** finds a line of best fit to the data. In the Algebra window **Author** the expression

$$\text{FIT}\big([\ell, m\ell + c], [20.0, 10.0], [38.5, 17.5], [48.0, 21.5], [57.0, 25.5], [67.0, 29.0]\big).$$

Now **Simplify** and **approX**, then DERIVE responds with the equation of the 'best' straight line through the data points. **Plot** this line to see how 'good' is the model. Figure 1.6 shows the screen print-out.

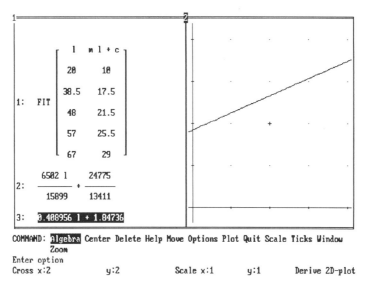

Figure 1.6   Linear model relating ℓ and T

The linear model relating T and ℓ is

$$T = 0.41\ell + 1.85 .$$

The data can be entered into DERIVE more efficiently than was shown in the solution to Example 1E.

In the Algebra window use the **Declare** and **Matrix** commands and select 5 rows and 2 columns. Enter the data in the order 20.0, 10.0, 38.5, 17.5, ... etc pressing enter after each number.

The data appears on the screen as an array of numbers with 5 rows and 2 columns. This array is called a *matrix*. The properties and algebra of matrices are introduced in Chapter 11. For the time being, you only need to think of a matrix as a convenient method of displaying a table of numbers.

Plot the matrix of data values to obtain a graph of the data.

To use the FIT function of DERIVE to find the equation of the line of best fit, highlight the matrix and **Author**

FIT($[\ell, m\ell + c]$, now press F3 to insert the data) .

Simplify to give the equation of the 'line of best fit'

$$0.408951\ell + 1.84735 .$$

Exercise 1C

1.   Use the FIT function in DERIVE to find the equation of the line of best fit for
      the following sets of data.

(a)	x	0.1	0.5	0.7	1.1	(b)	r	0.1	0.5	1.0	1.4	1.7	2.1
	y	1.1	2.6	3.6	5.1		s	−3.89	−2.65	−1.1	0.14	1.07	2.31

(c)	t	1	2	3	4	(d)	x	1.1	1.3	1.5	1.7	1.9
	v	10.2	0.4	−9.4	−19.2		p	2.20	2.55	2.89	3.23	3.58

2.   In an experiment the resistance of a length of wire was measured at different
      temperatures and the readings are shown in Table 1.4.

Table 1.4

Temperature T (°C)	50	80	100	120	160
Resistance R (ohms)	53.3	58.4	61.9	65.3	72.3

(a)   Draw a graph of the resistance against temperature and show that a
       linear law is a good model.

(b)   Find the equation of the line of best fit.

(c)   Use the model in part (b) to find the resistance of the wire at a
       temperature of 20°C.

3.   The speed of a car (in ms^{-1}) accelerating away from traffic lights is given at 0.5
      second time intervals in Table 1.5. From a linear model find the acceleration
      of the car.

Table 1.5

time t (seconds)	0.5	1.0	1.5	2.0	2.5	3.0
speed v (ms^{-1})	1.58	3.26	4.84	6.38	8.24	9.72

1.5   SOLVING LINEAR EQUATIONS

DERIVE ACTIVITY 1B

In this activity we will examine a procedure for solving linear equations.

(A)   (i)   **Author** the equation $4x - 2 = 10$.

     (ii)   Press A for **Author** followed by F4 and the original expression appears in brackets. Now type +2 and press enter. This will add 2 to both sides of the equation. **Simplify** the expression on the screen.

     (iii)   Again press A for **Author** followed by F4, but then type /4 and enter. This will divide both sides of the equation by 4. **Simplify** this expression to obtain $x = 3$.

     (iv)   Use the same approach to solve

$$5x - 6 = 14 \quad \text{and} \quad 3x - 2 = 19.$$

(B)   (i)   Solve the linear equations below using the **Author F4** approach as in (A), but this time subtract rather than add a number.

     (a) $5x + 6 = 26$ (c) $4x + 2 = 18$

     (b) $6x + 2 = 14$ (d) $6x + 7 = -5$

    (ii)   Now solve the equations below.

     (a) $6x + 4 = -2$ (d) $6x - \dfrac{1}{2} = \dfrac{7}{2}$

     (b) $\dfrac{x}{2} + 3 = 8$ (e) $0.31t + 7.12 = 5.3$

     (c) $\dfrac{x}{4} - 7 = 3$ (f) $3.06y - 4.17 = 11.24$

(C)   (i)   Sometimes the x will appear on both sides of the equals sign as in

$$5x + 12 = 7x + 4 .$$

Author this expression, then **Author F4** and type $-5x$ enter and **Simplify**. Now continue until you obtain $x = 4$.

(ii)   Solve the equations

(a) $5x + 2 = 12x - 54$
(b) $7x - 7 = 6x - 4$
(c) $9x - 11 = 5x - 4$
(d) $6x + 2 = 17x - 15.$
(e) $4.21x - 1.73 = 2.63x + 5.19$
(f) $13.03t + 4.82 = -0.61t - 2.17$

(D)   (i)   Consider now equations of the form

$$4(x+3) = 3 .$$

Author this equation and then **Expand** it. What happens to each term that was in the bracket?
Now complete the process of solving for x.

(ii)   Solve each of these equations starting with the **Expand** command.

(a) $4(x+7) = 8$ (c) $6(2x-5) = 15$

(b) $3(x-6) = 12$ (d) $7(3x-9) = 14$

(E)   Finally **Author** any of the equations above and press L for **soLve**.

Equations can be solved by carrying out the same operation on each side of the equals sign, to simplify the equation.

Forming and Solving Equations

Example 1F

When three consecutive numbers are added together their total is 144. What are these numbers?

Solution

Let x be the smallest of the three numbers, then $x + 1$ and $x + 2$ will be the other two. So their total is

$$x + x + 1 + x + 2 = 144$$

or $3x + 3 = 144$.

To solve this subtract 3 from both sides of the equation to give

$$3x = 141$$

and then divide by 3 to give

$$x = \frac{141}{3} = 47 \ .$$

So the lowest number is 47 and the others are 48 and 49.

Example 1G

The perimeter of a triangle is 27 cm. The longest side is 3 times as long as the shortest side and the other side is 8 cm shorter than the longest side. Find the length of each side.

Solution

Let x be the length of the shortest side, then the other two sides have lengths $3x$ and $3x - 8$. So the perimeter is given by

$$x + 3x + 3x - 8 = 27$$

Adding 8 to both sides gives

$$7x = 35 .$$

Dividing by 7 gives

$$x = 5 \text{ cm} .$$

So the sides are 5 cm, 15 cm and 7 cm.

Exercise 1D

1.   Without using DERIVE solve the equations

      (a) $6x - 18 = 12$ (g) $16 - 2x = x + 10$
      (b) $4x + 2 = 6x - 9$ (h) $4x + 2 = 8 - x$
      (c) $7x + 5 = 11$ (i) $7t - 11 = 2t + 9$
      (d) $6 - 3x = 4 + 2x$ (j) $3(t-4) = 2(t+7)$
      (e) $16x - 12 = 4$ (k) $0.21(x+1.2) = 0.37x$
      (f) $5x + 2 = -13$ (l) $1.46(x-0.9) = -4.1(2.61-2x).$

      (Check your answers using the **soLve** command of DERIVE).

2.   The sum of 3 consecutive odd numbers is 141. What are the numbers?

3.   The perimeter of a rectangle is 60 cm. Find the length and width if

      (a)   the length is twice the width,
      (b)   the length is 8 cm greater than the width.

4.   Using the formula $s = ut + \frac{1}{2}at^2$ find the unknown term if

      (a) $t = 1, s = 6, u = 5,$
      (b) $t = 2, s = 4, a = -2,$
      (c) $t = 10, s = 5, a = 0.5.$

1.6   SIMULTANEOUS EQUATIONS

Often two equations may apply at the same time, for example $y = 4x + 6$ and $y = 3x + 10$. These are known as *simultaneous equations*. Is there a pair of values for x and y which will satisfy both equations? If so how do we find them?

DERIVE ACTIVITY 1C

(A)   (i)   Suppose that we have the two equations above. In DERIVE **Author** and **Plot** both of them. To obtain a good plot use the **Scale** command with x:5 and y:5. **Move** the cross to (5,10) and **Centre**.

(ii)   Use the arrow keys to move the cross to the point where the lines cross. This should have an x value of about 4 and y value of about 22.

(iii)   Return to the Algebra window. Highlight the first equation ($y = 4x + 6$) and press **Manage** and **Substitute**, do not enter a value for x, but for y, type $3x + 10$. Now use **soLve** to find the value of x in the equation you obtain. Now highlight either of the original equations, and use **Manage Substitute**, giving the value 4 for x. Then use **Simplify** to obtain the y value.

(B)   Use the approach above to solve the following pairs of equations; adjusting the scales of your plot as necessary.

(i) $y = 20 - 5x$ (iii) $y = 10 - x$
       $y = 4 + 3x$ $y = 4 + 3x$

(ii) $y = 2x + 4$ (iv) $y = 7 - 2x$
       $y = x + 10$ $y = 4 + 3x$

(C)   (i)   Often simultaneous equations are presented in the form $4x + 2y = 8$ and $6x - 5y = 10$. **Author** and try to **Plot** the first of these equations. Return to the Algebra window and use **soLve**, selecting y as the variable. Now you can plot this equation. Use **soLve** to help you plot the second equation. With the cursor try to read off the values of x and y where the lines cross.

(ii)   Highlight the first equation and use **Manage Substitute**. Do not enter a
        variable for x. For y first delete the y on the command line, now move
        the highlight to your last expression, the rearranged form of the second
        equation, press the right arrow twice and the right hand side will be
        highlighted on its own. Now press F3, notice that the whole expression
        appears, and then enter. Use **soLve** on the equation you obtain. Try a
        similar approach to find y by substituting the value of x back into one of
        the original equations.

(D)   Use the approach of (C) to solve the simultaneous equations below.

(i) $x + y = 6$ (iii) $2x + 3y = 5$
        $x - y = 7$ $3x - 2y = 4$

(ii) $2x + 3y = 6$ (iv) $x + y = 6$
        $5x - 2y = 10$ $2x - y = 7$

(E)   Finally **Author** $[2x + 3y = 6, 5x - 2y = 10]$ and **soLve**.

Pairs of simultaneous equations can be solved using the substitution approach
described. However, in DERIVE the pair of equations

$$ax + by = c \quad \text{and} \quad dx + ey = f$$

can be solved directly by using the commands **Author** $[ax + by = c, dx + ey = f]$ and
soLve.

(F)   Solve the simultaneous equations in (D) directly by using DERIVE.

Solving Simultaneous Equations

The method of solving simultaneous equations in DERIVE Activity 1C was to substitute the formula for y from one equation into the other. This method works well but can involve some tedious algebra. An alternative approach when solving simultaneous equations by hand is to add multiples of the two equations together to eliminate one variable. This is illustrated in the examples below.

Example 1H

Solve the pair of simultaneous equations

$$3x + 2y = 6$$
$$x + 2y = 4 .$$

Solution

First it is helpful to label the equations (1) and (2) as below.

$$3x + 2y = 6 \qquad (1)$$
$$x + 2y = 4 \qquad (2)$$

Here both equations contain the same number of y's so equation (2) can be subtracted from equation (1) to give

$$(3x - x) + (2y - 2y) = 6 - 4$$
$$2x \qquad\qquad = 2$$

So $x = 1$. Substituting this into the first of the original equations gives

$$3 + 2y = 6$$
$$2y = 3$$
$$y = 1.5 .$$

So the solution is $x = 1$ and $y = 1.5$.

Example 1I

Solve the pair of simultaneous equations

$$3x - 4y = 2$$
$$2x + 3y = 6 \,.$$

Solution

Labelling the equations (1) and (2) is helpful,

$$3x - 4y = 2 \qquad (1)$$
$$2x + 3y = 7 \qquad (2)$$

As a first step multiply equation (1) by 2 and equation (2) by 3. This will mean that both equations end up with the same number of x's in each.

$$6x - 8y = \ 4 \qquad 2 \times (1)$$
$$6x + 9y = 21 \qquad 3 \times (2)$$

Now subtracting the second of these from the first gives

$$(6x - 6x) + (-8y - 9y) = 4 - 21$$
$$-17y = -17$$
$$y = 1 \,.$$

Substituting this value of y into the first equation and solving gives $x = 2$. So the solution is $x = 2$ and $y = 1$.

Exercise 1E

1.   Solve each of the following pairs of simultaneous equations 'by hand'.

   (a) $x + y = 16$ (d) $7x - 2y = 18$
         $x - y = \ 4$ $3x + 4y = \ 2$

   (b) $2x + 3y = 12$ (e) $4a + 5b = \ 6$
         $4x + 5y = 22$ $3a - 2b = 16$

   (c) $4x + 3y = 30$ (f) $3u + 6t = 24$
         $5x - 2y = 26$ $4u + 5t = 17$

   (Check your answers using DERIVE).

1.7   QUADRATICS

In this section we introduce another rule between two variables x and y which uses the term x^2 as well as x. The rule is called a *quadratic law* and has the general form
$y = ax^2 + bx + c$. For example $y = x^2 + 2x - 3$ is a quadratic law.

DERIVE ACTIVITY 1D

(A)   Many physical quantities are related by quadratic laws. As a stone falls the distance that it has travelled after certain times are given in the Table 1.6.

Table 1.6

time (seconds)	0	1	2	3	4
distance fallen (m)	0	5	20	45	80

(i)   Use the **Declare** and **Matrix** commands and select 5 rows and 2 columns and enter the data as a matrix. (Remember to enter the data in the order 0, 0, 1, 5, 2, 20 ... etc.

Before plotting use **Scale** to select x:1 and y:20, **Move** to (2,40) and **Centre**. Now **Plot** the data.

(ii) Return to the Algebra window. It is possible to find the equation of a quadratic curve that passes through the points. **Author**

$$\text{FIT}([x, ax^2 + bx + c], \text{ now press F3 to insert the data}).$$

Then **Simplify** and **Plot**, to obtain the equation of a curve of best fit that passes through all these points and its graph. Figure 1.7 shows a DERIVE screen dump of this activity.

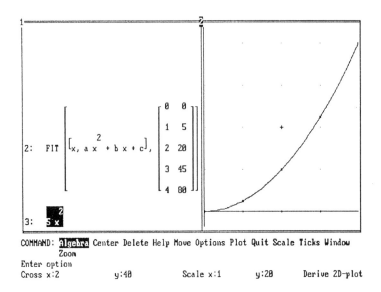

Figure 1.7

(B)   The Highway Code gives the following table of speed and stopping distances.

Table 1.7

Speed (ms^{-1})	8.9	13.3	17.8	22.2	26.7	31.1
Stopping Distance (m)	12	23	36	53	73	96

Enter the data into a matrix and plot the points. Then use the FIT command to find the equation of the quadratic curve through these points, plotting the result.

(Note that in the Highway Code the speeds are given in mph, they have been converted to ms^{-1} using the rule 1 mph $\equiv$ 0.444 ms^{-1}).

(C)   (i)   **Remove** all your expressions and **Delete All** your plots. Change the
             Scale to $x:1$, $y:1$, **Move** the cross to $(0,0)$ and **Centre**. Now **Author** and
             Plot the expressions.

> (a) $y = x^2$ (c) $y = x^2 + 2$
>
> (b) $y = x^2 + 1$ (d) $y = x^2 - 1$

Describe how each curve is related to $y = x^2$.
Delete All your plots.

(ii)   Change the **Scale** to $x:2$ and $y:1$. **Author** and **Plot** each of the
       expressions below.

> (a) $y = x^2$ (c) $y = (x - 2)^2$
>
> (b) $y = (x - 1)^2$ (d) $y = (x+2)^2$

Describe how each curve is related to $y = x^2$.
Delete All your plots.

(iii)   **Author** each of the expressions below. Try to predict the coordinates of
        the lowest point of the curve. Plot to check your predictions.

> (a) $y = x^2 + 1$ (d) $y = (x+2)^2 + 4$
>
> (b) $y = (x+1)^2$ (e) $y = -x^2$
>
> (c) $y = (x+1)^2 - 1$ (f) $y = 4 - x^2$

This activity (C) shows that all quadratics are similar in shape to the basic
quadratic $y = x^2$. The effect of the 'other numbers' is to translate the basic shape of the
graph of $y = x^2$.

The curve $y = x^2 + a$ is translated up a units (if a is positive) and has a minimum
value of a when $x = 0$.

The curve $y = (x-b)^2$ is translated to the right by b units (if b is positive) and has a
minimum value of zero when $x = b$.

The curve $y = (x-b)^2 + a$ is translated to the right by b units and up by a units and has
a minimum value of a when $x = b$.

Solving Quadratic Equations

A quadratic equation is an equation of the form

$$x^2 + 5x + 6 = 0 .$$

The approaches that you have used for solving linear equations are not appropriate here. The DERIVE Activity 1E leads you into some of the important ideas needed for solving quadratic equations.

DERIVE ACTIVITY 1E

(A) (i) **Author** and **Plot** $x^2 + 5x + 6$. You will see that the curve crosses the x-axis twice and so has 2 solutions. Return to the Algebra window and use **soLve** to find the value of these 2 solutions.

(To obtain two solutions you should set your **Options** to **Precision Mixed**).

(ii) Repeat for $x^2 + 6x + 9$. This time you will see that there is only one solution.

(iii) Repeat for $x^2 + 5x + 10$. This time you will see that the curve does not cross the axis and there are no real solutions. Using **soLve** will produce strange looking expressions involving î. These are known as *complex numbers* and will be dealt with in Chapter 10.

(B) (i) Highlight $x^2 + 5x + 6$ and **Factor**, selecting **Rational**. What is the connection between the new expression and the solutions?

(ii) Repeat for $x^2 + 6x + 9$.

(iii) What happens for $x^2 + 5x + 10$?

(C) (i) **Author** the expression $(x+a)(x+b)$ and **Expand**, responding to **EXPAND variable 1:** with x. Explain how

$$(x+a)(x+b) \text{ is related to } x^2 + px + q.$$

Write down p and q in terms of a and b.

(ii)   Write down what you would expect to get when you **Expand** each of the
         expressions below and then check your answer on DERIVE.

         (a) $(x+2)(x+4)$
         (b) $(x+1)(x+5)$
         (c) $(x+3)(x-4)$
         (d) $(x-6)(x-2)$

         Now **Plot** each expression and find the solutions of the quadratic
         equation formed when each expression is put equal to zero.

(iii)   When $x^2 + px + q$ is factored it produces the two brackets $(x+a)(x+b)$
         where $ab = q$ and $a + b = p$. For example with

$$x^2 + 7x + 10$$

         we have $ab = 10$ and $a + b = 7$. Clearly these are satisfied by $a = 5$ and
         $b = 2$, so

$$x^2 + 7x + 10 = (x+5)(x+2) .$$

         Try to factorise each expression below, checking your answers with
         DERIVE.

         (a) $x^2 + 5x + 4$ (d) $x^2 - 5x + 6$
         (b) $x^2 + 6x + 8$ (e) $x^2 - x - 6$
         (c) $x^2 + 3x + 2$ (f) $x^2 + x - 6$

A quadratic equation can have

(i)   two roots if it crosses the x-axis twice,
(ii)   one root if it just touches the x-axis,
(iii)   two complex roots if it does not cross the x-axis.

The quadratic equation $x^2 + px + q$ has factors $(x+a)$ and $(x+b)$ where
$a + b = p$ and $ab = q$.

Solving Quadratics by Factorisation

It is possible to solve quadratics by factorisation, an approach which is demonstrated in the examples below.

Example 1J

Solve $x^2 + 7x + 12 = 0$ by factorisation.

Solution

We have to write $x^2 + 7x + 12 = 0$ in the form $(x+a)(x+b) = 0$.

So $a + b = 7$ and $ab = 12$. Possible values of a and b that have a product of 12 are 1 and 12, 2 and 6, 3 and 4. But a and b must also have a sum of 7 and so $a = 3$ and $b = 4$, giving

$$(x+3)(x+4) = 0 .$$

The product of these brackets will be zero when either bracket is zero ie. when $x + 3 = 0$ or $x + 4 = 0$ giving $x = -3$ and $x = -4$ as the solutions of the equation.

Example 1K

Solve $3x^2 + 14x + 8 = 0$ by factorisation.

Solution

The equation $3x^2 + 14x + 8 = 0$ must be written in the form $(3x+a)(x+b) = 0$. So $ab = 8$ and $a + 3b = 14$. Possible values of a and b that have a product of 8 are 1 and 8 or 2 and 4. To satisfy $a + 3b = 14$ we must use $a = 2$ and $b = 4$ giving

$$(3x+2)(x+4) = 0 .$$

The product of these brackets will be zero when either is zero giving

$$3x + 2 = 0 \quad \text{or} \quad x + 4 = 0$$
$$x = -\tfrac{2}{3} \quad \text{or} \quad x = -4$$

So the solutions of the equation are $x = -\tfrac{2}{3}$ and $x = -4$.

Solving Quadratics with the Formula

Often quadratic equations are difficult or impossible to solve using the technique of factorisation. However there is a formula to solve the quadratic equation $ax^2 + bx + c = 0$. It is given by

$$x = \frac{-b \pm \sqrt{b^2 - 4ac}}{2a} \, .$$

When using this formula, note that

(i)   if $b^2 - 4ac > 0$ then there will be 2 solutions,

(ii)   if $b^2 - 4ac = 0$ then there will be only 1 solution,

(iii)   if $b^2 - 4ac < 0$ there will be no real solutions as the square root of a negative number is not defined (until Chapter 10).

[We will not prove the formula here. To see a derivation you should refer to another text].

Example 1L

Find the solutions of

$$3x^2 + 4x - 8 = 0 \, .$$

Solution

Here

$$a = 3, \quad b = 4 \quad \text{and} \quad c = -8.$$

Substituting into the formula gives

$$x = \frac{-4 \pm \sqrt{4^2 - 4 \times 3 \times (-8)}}{2 \times 3} = \frac{-4 \pm \sqrt{16 + 96}}{6} = \frac{-4 \pm \sqrt{112}}{6}$$

Hence $x = 1.10$ or -2.43 (to 3sf).

Exercise 1F

1.   Sketch the following quadratics.

   (a) $y = x^2 + 6$ (d) $y = 2 - (x+3)^2$

   (b) $y = (x+4)^2$ (e) $y = (x+3)^2$

   (c) $y = 6 - x^2$ (f) $y = (x-4)^2 + 2$

2.   Factorise the following quadratics.

   (a) $x^2 + 9x + 20$ (f) $2x^2 - 9x - 5$

   (b) $x^2 - 7x + 12$ (g) $2x^2 + 5x + 2$

   (c) $x^2 + 4x + 3$ (h) $5x^2 - 34x - 7$

   (d) $x^2 + 5x + 6$ (i) $3x^2 - 10x - 8$

   (e) $x^2 - 1$ (j) $x^2 - 4$

3.   Solve the following quadratic equations by factorisation.

   (a) $x^2 + 6x + 9 = 0$ (c) $2x^2 + x - 10 = 0$

   (b) $x^2 + 10x + 16 = 0$ (d) $x^2 - 16 = 0$

4.   Find the solutions, if any, of the following quadratic equations.

   (a) $x^2 + 6x - 7 = 0$ (e) $6x^2 - 10x + 2 = 0$

   (b) $3x^2 + 4x - 9 = 0$ (f) $5x^2 - 9x + 8 = 0$

   (c) $x^2 + 8x + 20 = 0$ (g) $13x^2 + 81x - 94 = 0$

   (d) $3x^2 - 4.5x + 9 = 0$ (h) $6x^2 - 42 = 0$

1.8   POLYNOMIALS

The linear and quadratic functions considered so far in this chapter are simple examples of *polynomials*. A polynomial is any function of the form

$$a_0 + a_1x + a_2x^2 + a_3x^3 + \dots + a_nx^n .$$

The numbers $a_0, a_1, a_2, \dots, a_n$ are referred to as the *coefficients* of the polynomial and n the largest power is called the *degree* of the polynomial.

DERIVE ACTIVITY 1F

(A)   **Author** and **Plot** the following functions.

  (i) x^2 (ii) x^3 (iii) x^4 (iv) x^5 (v) x^6 (vi) x^7 (vii) $-x^3$ (viii) $-x^4$

  What do you notice? **Delete All** your plots.

(B)   **Author** and **Plot** the following.

  (i) x^3 (ii) $x^3 + 2$ (iii) $x^3 - x$ (iv) $x^3 - x^2$

  How many solutions can a polynomial of degree 3 have? Is there a minimum number of solutions?

(C)   Investigate how many times a polynomial of degree 4 crosses the x-axis.

  By choosing polynomials of different degree show that the number of roots (and factors) of a polynomial is less than or equal to the degree.

(D)   It is possible to express polynomials in the form

$$(x-a)(x-b)(x-c) \dots$$

  Try Authoring and Plotting for different numbers of brackets and different values in the brackets. What do you notice? What happens if some brackets are repeated?

Exercise 1G

Sketch each of the following polynomials.

(a) $y = x^5 + 1$ (b) $y = x^6 - 1$ (c) $y = (x+1)^4 - 1$
(d) $y = (x+1)(x-1)(x+2)$ (e) $y = (x+1)^2(x-1)$ (f) $y = x(x-2)^2$

1.9 FUNCTIONS

In earlier sections we have introduced the idea of variables being related to each other by rules or laws, for example, the tension in an elastic string T is linearly related to its length, ℓ. The measured quantities T and ℓ are called *variables* because their values vary in an experiment to find the linear law. In such an experiment we would probably choose the tension and then measure the corresponding length. The chosen variable is called the *independent variable* and the other variable is called the *dependent variable* because its value depends on the value chosen for the independent variable.

When the two variables x and y say are related so that the one quantity y depends on the value of the other x, then y is said to be a *function* of x. The only restriction on such a relationship being called a function is that for each possible value of x there is only one value of y. An example of a quadratic function is

$$y = x^2 + 6 \quad \text{for} \quad x \geq 0.$$

Notice that the independent variable is restricted to positive values of x. This is called the *domain* of the function. The values of y are restricted to $y \geq 6$ and this set of values is called the *range* of the function.

It is often convenient to use a special notation for functions. For example, instead of writing

$$y = x^2 + 6$$

we could write

$$f(x) = x^2 + 6 .$$

The value of a function for a particular value of x is written in a special way. For example, suppose that we want the value of $f(x)$ for $x = 5$. Then we denote this by $f(5) = 5^2 + 6 = 31$. Similarly $f(2) = 2^2 + 6 = 10$ is the value of $f(x)$ for $x = 2$.

Different letters can be used to define different functions, for example

$$g(x) = \frac{x}{2} \quad \text{and} \quad b(x) = x + 5.$$

DERIVE ACTIVITY 1G

(A)   It is a simple matter to define functions in DERIVE.

   (i)   Suppose we want to define the function $f(x) = 5x$. Respond to the
          requests of DERIVE using **Declare**, **Function** in the following way:

                    DECLARE FUNCTION name: f
                    DECLARE FUNCTION value: $5x$

          DERIVE then assigns the rule

                    $F(x): = 5x$.

   (ii)   Repeat the process to define $g(x) = x + 2$ and $h(x) = x/2$.

(B)   (i)   Author $f(6)$ and **Simplify**.

   (ii)   Find $h(2)$, $g(7)$ and $f(3)$ in the same way.

(C)   (i)   Now **Author** $f(g(x))$ and **Simplify**.
          Explain what has happened.

   (ii)   Repeat for $g(f(x))$, $h(g(x))$ and $g(h(x))$.

   (iii) Write down what you would expect to obtain for $f(h(x))$ and $h(f(x))$.

(D)   (i)   Define further functions $s(x) = x^2$ and $r(x) = \sqrt{x}$.

   (ii)   **Author** and **Simplify** $s(g(x))$ and $r(g(x))$.

   (iii) Write down what you would expect for $s(r(x))$ and $r(s(x))$. Check your
          predictions with DERIVE.

Composite Functions

When two functions are combined as in DERIVE Activity 1G a *composite function* is formed. The rule is that the functions are applied in order from the right, so for $fg(x)$, g is applied to x, and then the rule of f is applied to the result.

Example 1M

If $f(x) = x + 5$, $g(x) = x^2$ and $h(x) = x - 7$, find $fg(x)$, $gf(x)$ and $gh(x)$.

Solution

To find $fg(x)$ first note that $g(x)$ gives x^2. So

$$f(g(x)) = f(x^2) = x^2 + 5.$$

To find $gf(x)$ first note that $f(x)$ gives $x + 5$. So

$$g(f(x)) = g(x+5) = (x+5)^2.$$

To find $g(h(x))$ first note that $h(x)$ gives $x - 7$. So

$$g(h(x)) = g(x-7) = (x-7)^2.$$

Exercise 1H

1.   If $f(x) = x^2$, $g(x) = \dfrac{x}{2}$ and $h(x) = x - 6$, find

   (a) $f(0)$ (d) $f(2)$

   (b) $g(2)$ (e) $g(10)$

   (c) $h(9)$ (f) $h(-1)$.

2.   For f, g and h as defined in problem 1, find

   (a) $fg(x)$ (d) $hg(x)$

   (b) $gh(x)$ (e) $fgh(x)$

   (c) $gf(x)$ (f) $ghf(x)$.

Inverse Functions

The *inverse of a function* will reverse the effect of the original function. If $f(x)$ is a function its inverse is written as $f^{-1}(x)$.

For example if $f(x) = x + 5$ then $f^{-1}(x) = x - 5$. Notice what happens to a value of x under f and then apply f^{-1} to the answer $f(x)$. For $x = 2$, $f(2) = 7$ and then $f^{-1}(7) = 2$. So $f^{-1}(f(2)) = 2$.

As another example, let $g(x) = \dfrac{x}{2}$ then $g^{-1}(x) = 2x$. If $x = 4$, then $g(4) = 2$ and $g^{-1}(2) = 4$ so that $g^{-1}(g(2)) = 2$.

Often functions are not this simple. Consider,

$$h(x) = \frac{(x-2)^2}{5} \ .$$

$h(x)$ could be represented by a flow chart as below.

$$x \quad \rightarrow \quad \boxed{-2} \quad \rightarrow \quad \boxed{\text{SQUARE}} \quad \rightarrow \quad \boxed{\div 5} \quad \rightarrow \quad h(x) = \frac{(x-2)^2}{5}$$

The inverse can be obtained by reversing the flow chart and using the inverse of the individual functions as shownbelow.

$$h^{-1}(x) = \sqrt{(5x)} + 2 \quad \leftarrow \quad \boxed{+2} \quad \leftarrow \quad \boxed{\text{SQUAREROOT}} \quad \leftarrow \quad \boxed{\times 5} \quad \leftarrow \quad x$$

Forming the composite of h and h^{-1} gives

$$h^{-1}(h(x)) = h^{-1}\left(\frac{(x-2)^2}{5}\right) = \sqrt{\frac{5(x-2)^2}{5}} + 2$$

$$= x \ .$$

In general if the inverse of a function $f(x)$ is denoted by $f^{-1}(x)$ then

$$f^{-1}(f(x)) = x \ \text{ and } \ f(f^{-1}(x)) = x.$$

DERIVE ACTIVITY 1H

(A)   (i)   **Author** x^2 and $\sqrt{x}$, a function and its inverse (use Alt-Q to get $\sqrt{}$).
              Plot both functions. What is the relationship between the graph of the
              function and its inverse? (Hint: plot $y = x$).

       (ii)   Repeat for $5x$ and $x/5$.

       (iii) Repeat for $x + 1$ and $x - 1$.

(B)   (i)   To obtain the inverse of functions, such as $f(x) = 2x + 1$,
              Author $y = 2x + 1$ and then **soLve** for x. This will produce the inverse
              function $f^{-1}(y)$, ie. using y instead of x. However this makes no
              difference when plotting in DERIVE.

       (ii)   Repeat the process above for each function below.

              (a) $f(x) = 6x + 2$ (d) $f(x) = \dfrac{1}{4 - x}$

              (b) $f(x) = \dfrac{1}{x + 2}$ (e) $f(x) = \dfrac{1}{x - 3}$

              (c) $f(x) = \dfrac{x + 2}{x - 1}$ (f) $f(x) = \sqrt{(5 - x)}$

       (iii) Now try finding and plotting the inverse functions for each function
              below.

              (a) $f(x) = x^2 + 8$ (c) $f(x) = 4(x - 7)^2$

              (b) $f(x) = x^2 - 8$ (d) $f(x) = \dfrac{x^2}{1 - x^2}$

              Explain why two expressions are given when you used the **soLve**
              command.

(C)   (i)   **Remove** all your expressions from the Algebra window. Use **Plot Window Close** to remove the plot window.

(ii)   **Author** the expression $y = 6x - 7$. Now press **Author** and F4 followed by +7. **Simplify** your result, noting that 7 has been added to both sides of the equation. Now use **Author** and F4 again this time followed by /6. **Simplify** your result, noting that both sides of the equation have been divided by 6. You should now be able to see that the inverse of the

original function $f(x) = 6x - 7$ is given by $f^{-1}(x) = \dfrac{x+7}{6}$, when the x's and y's are interchanged.

(iii)   Now to find the inverse of $f(x) = \dfrac{1}{x+1}$. First **Author** $y = \dfrac{1}{x+1}$.
Use **Author** and F4 followed by $(x+1)$.
Simplify and then use **Author** and F4, followed by /y.
Simplify and then use **Author** and F4, followed by –1.
Simplify your expression to show that

$$f^{-1}(x) = \frac{1}{x} - 1,$$

when the x's and y's are interchanged.

(iv)   Use the approach described above to find the inverse of each function below.

(a) $f(x) = 7x + 4$ (e) $f(x) = \sqrt{x-5}$

(b) $f(x) = \dfrac{1}{5x-3}$ (f) $f(x) = \dfrac{5(x-6)}{7}$

(c) $f(x) = 4x - 7$ (g) $f(x) = \dfrac{18(x-5)}{5}$

(d) $f(x) = \sqrt{9-x}$ (h) $f(x) = \dfrac{x+1}{x-2}$

Finding Inverse Functions

It is possible to find some inverse functions by using flow charts, but this method will not always work. The method demonstrated in DERIVE must then be used as shown in the examples below.

Example 1N

Find the inverse of the function

$$f(x) = \frac{x}{x-2} .$$

Solution

First write the function in the form

$$y = \frac{x}{x-2} .$$

Now this equation must be rearranged in the form $x = ...$

First multiply both sides by $(x-2)$.

$$(x-2)y = \frac{x(x-2)}{(x-2)}$$
$$xy - 2y = x .$$

Now taking all the terms that contain x to the left hand side (LHS) and those that do not contain x to the right hand side (RHS) gives

$$xy - x = 2y$$
$$x(y-1) = 2y .$$

Now dividing by $(y-1)$ gives

$$x = \frac{2y}{y-1} .$$

So the inverse function is given by

$$f^{-1}(x) = \frac{2x}{x-1} .$$

Example 1P

Find the inverse of the function

$$g(x) = \frac{x+3}{x-1} .$$

Solution

First the function is written in the form

$$y = \frac{x+3}{x-1} .$$

Then it must be rearranged to make x the subject, ie. in the form $x = \ldots.$

First multiply both sides by $(x-1)$ to give

$$y(x-1) = x + 3$$

or

$$xy - y = x + 3 .$$

Then take all the terms that contain x to the LHS and all those that do not contain x to the RHS giving

$$xy - x = y + 3$$
or $$x(y-1) = y + 3 .$$

Now dividing by $(y-1)$ gives

$$x = \frac{y+3}{y-1} .$$

Now the inverse function can be written as

$$g^{-1}(x) = \frac{x+3}{x-1} .$$

Exercise 1I

1.   Find the inverse of each function below by drawing flow charts.

   (a) $f(x) = 6x - 10$ (e) $f(x) = \sqrt{2x - 4}$

   (b) $f(x) = 4(x+2)$ (f) $f(x) = 4\sqrt{(x+1)} - 5$

   (c) $f(x) = \sqrt{(2x+7)}$ (g) $f(x) = \sqrt{\left(\dfrac{x}{2} + 1\right)} - 5$

   (d) $f(x) = 8x^3 - 5$ (h) $f(x) = \dfrac{\sqrt{x^3 + 4}}{5}$

2.   Find the inverse of each function below.

   (a) $f(x) = \dfrac{x-4}{x+2}$ (d) $f(x) = \dfrac{1}{x} + \dfrac{1}{2x}$

   (b) $f(x) = \dfrac{x}{1+x}$ (e) $f(x) = \dfrac{x-6}{x+2}$

   (c) $f(x) = \dfrac{x^3}{x^3 + 2}$ (f) $f(x) = \dfrac{x^3 + 3}{2x^3 + 1}$

3.   The functions f, g and h are defined as $f(x) = x^3$, $g(x) = \dfrac{1}{x+6}$ and $h(x) = 2x - 5$. Find

   (a) $f^{-1}(0)$ (d) $g^{-1}(8)$

   (b) $f^{-1}g(2)$ (e) $g^{-1}(-2)$

   (c) $h^{-1}(4)$ (f) $h^{-1}(0)$.

1.10   DISCONTINUOUS FUNCTIONS

Functions can be formed by fractions where the denominator can be zero for certain values. For example,

$$y = \frac{6}{x-1}$$

will have a zero denominator when $x = 1$, and then y will be infinite. The value of $x = 1$ is called a *discontinuity*. The DERIVE Activity 1I will help you to discover the properties of some of these functions.

DERIVE ACTIVITY 1I

(A)   (i)   **Author** and **Plot** $\dfrac{1}{x}$, selecting **Scales** x:2 and y:2.

At what value of x does the function have its discontinuity?
What happens to the function for large positive or negative x values?

(ii)   Repeat (i) for each function below.

(a) $\dfrac{1}{x+1}$ (b) $\dfrac{1}{x-1}$ (c) $\dfrac{1}{x+2}$

(iii)   For what values of x will the functions below have discontinuities?

(a) $\dfrac{1}{x-4}$ (b) $\dfrac{1}{x+10}$ (c) $\dfrac{1}{x-3}$

Check your answers by plotting with suitable scales.

(B)   (i)   **Remove** all the expressions in the **Algebra** window and **Delete All** your plots in the **Plot** window.

(ii)   **Author** and **Plot**

$$\frac{1}{(x-1)(x+2)}$$

selecting **Scales** x:2 and y:2. Where is this function discontinuous?
What happens to the function for large positive or negative values of x?
For what range of values is the function negative?

(iii)   Repeat (ii) for each function below. You may need to select different **Scales**.

(a) $\dfrac{1}{(x+1)(x-3)}$ (c) $\dfrac{1}{(2x-5)(3x-2)}$

(b) $\dfrac{1}{(x-2)(x-4)}$ (d) $\dfrac{1}{x(x-1)}$

(iv)   Predict where each function below will have discontinuities.

(a) $\dfrac{1}{(x+1)(x-7)}$ (c) $\dfrac{1}{x(x-4)}$

(b) $\dfrac{1}{(2x+3)(x+4)}$ (d) $\dfrac{1}{(2x-5)(x+2)}$

Check your answers by plotting each function. **Delete All** your graphs and **Remove** all your expressions.

(C)   (i)   **Author** and **Plot**

$$\frac{x+1}{x-2}$$

selecting **Scales** x:5 and y:5. Where is there a discontinuity of this function? What happens to the function for large positive or negative values of x? (**Author** and **Plot** the function of $y = 1$).

It appears that the function gets closer and closer to 1 for large values of x. This can be checked using the **Calculus Limit** commands, selecting variable: x and point: inf. Then **Simplify** the expression you obtain. The limit gives the value that the function approaches but never quite reaches as x gets bigger and bigger. The line $y = 1$ is called an *asymptote*.

(ii)   Repeat (i) for the functions below.

(a) $\dfrac{2x+4}{x-1}$ (c) $\dfrac{(x-3)(x+4)}{(x+1)(2x-4)}$

(b) $\dfrac{3x-5}{6x+7}$ (d) $\dfrac{(x-1)(x+1)}{(2x-1)(2x+1)}$

Summary

DERIVE Activity 1I has introduced the idea of an asymptote and shown graphs of
functions with discontinuities. To find the discontinuities we set the denominator to
zero and solve for x.

To find any horizontal asymptotes we introduce the idea of the limit of the function as
x gradually increases to $+\infty$ or gradually decreases to $-\infty$. For example, consider the

function $f(x) = \dfrac{(x+1)^2}{(2x-3)(x+2)}$.

Table 1.8 illustrates what happens as x increases.

Table 1.8

x	$f(x)$
1	−1.333
10	0.593
100	0.508
1000	0.501

As x grows larger and larger, the value of the function approaches the value 0.5. This
is written as

$f(x) \to 0.5$ as $x \to \infty$.

Similarly Table 1.9 shows what happens as x approaches $-\infty$.

Table 1.9

x	$f(x)$
−1	0
−10	0.440
−100	0.493
−1000	0.499

We deduce

$f(x) \to 0.5$ as $x \to -\infty$ and that $y = 0.5$ is a horizontal asymptote.

The vertical asymptotes are $x = 1.5$ and $x = -2$. Figure 1.8 shows a sketch of the function between the asymptotes.

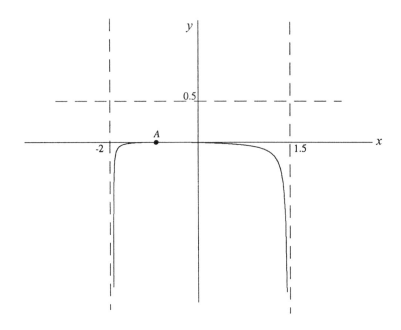

Figure 1.8 Graph of $f(x) = \dfrac{(x+1)^2}{(2x-3)(x+2)}$

In Chapter 6 you will learn how to sketch graphs of such functions by finding other properties of functions such as point A in Figure 1.8 which is called *a turning point*.

Exercise 1J

Find any horizontal and vertical asymptotes of the following functions. Use DERIVE to obtain graphs of the functions to confirm your answers.

(a) $y = \dfrac{3}{x-1}$

(b) $y = \dfrac{6}{3x+2}$

(c) $y = \dfrac{1}{x-1} + \dfrac{1}{x-2}$

(d) $y = \dfrac{1}{x} + \dfrac{1}{x+2}$

(e) $y = \dfrac{2x}{x+1}$

(f) $y = \dfrac{x}{x-4}$

(g) $y = \dfrac{3}{(x+2)(x-4)}$

(h) $y = \dfrac{x^2}{(x-5)(x+4)}$

2

Exponential and logarithmic functions

2.1 INTRODUCTION

In Chapter 1 you have seen the properties of linear and polynomial functions and how these functions can be used to model physical situations. For example, $T = ke$ models the tension in an elastic spring for an extension e, and $s = \frac{1}{2}gt^2$ is the distance travelled by a ball when dropped vertically as a function of time t. In this chapter we explore other power laws.

As an example consider the motion of a simple pendulum consisting of a small heavy object attached to an inextensible string. Table 2.1 shows the period of the pendulum for different string lengths.

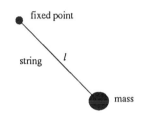

Figure 2.1   A simple pendulum

Table 2.1

Length ℓ (m)	0.5	0.6	0.7	0.8	0.9	1.0
Period T (s)	1.42	1.55	1.68	1.80	1.90	2.01

Figure 2.2 shows a graph of the period against the length. Clearly the graph is not linear and it does not have the general shape of a positive power of ℓ, examples of which are shown in Figure 2.3.

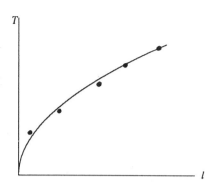

Figure 2.2

A graph of T against ℓ

Figure 2.3

Graphs of $y = x$, $y = x^2$ and $y = x^3$

In fact Figure 2.2 is a graph of the form $\ell^{0.5}$ or $\sqrt{\ell}$. The actual model relating T and ℓ is

$$T = 2.01\sqrt{\ell} \quad \text{or} \quad T = 2.01\ell^{0.5}.$$

A different type of power law occurs for the growth of money invested in a savings account. Suppose that you invest £3000 at 7% per annum. The growth of your investment is shown in Table 2.2.

Table 2.2

Time (years)	1	2	3	4	5
Amount (£)	3210	3434.7	3675.13	3932.39	4207.66

The model of this situation is

amount $= 3000(1.07)^t$.

Notice that in this model the variable itself (ie. t) is the power. This is often the situation in growth and decay processes.

The need to manipulate powers of ten occurs often in science and engineering because of the standard form of representing large and small numbers. For example, the mass of the earth is written as 5.98×10^{24} kg and the mass of an electron is 9.1×10^{-31} kg. This is certainly more convenient than writing 0.000 000 000 000 000 000 000 000 000 000 91 kg.

Each of these examples is of the form a^n which is called an *index law* or *exponential law*; a is called the *base* and n is called the *exponent* or *index* or *power*.

MINI-INVESTIGATION

1.   Using DERIVE **Author** and **Simplify** the expressions below

   (a) $(x^2)(x^3)$ (b) $(x^6)(x^4)$ (c) $(x^m)(x^n)$

   What do you notice? Now try

   (d) $(x^4)(y^4)$ (e) $(x^7)(b^7)$ (f) $(z^2)(x^2)$

   What do you notice?

2.   **Remove** your expressions. **Author** and **Simplify**

   (a) $(x^8) \div (x^5)$ (b) $(x^6) \div (x^8)$ (c) $(z^6) \div (z^4)$

   (d) $(x^3) \div (x^9)$ (e) $(x^6) \div (z^6)$ (f) $(z^4) \div (a^4)$

   What do you notice?

3.   **Remove** your expressions. **Author** and **Simplify**

   (a) $(x^2)^4$ (b) $(x^3)^6$ (c) $(x^3)^4$ (d) $(x^n)^m$

   What do you notice?

4.   **Remove** your expressions. **Author** and **Simplify**

   (a) 2^0 (b) x^0 (c) $x^{\frac{1}{2}}$

   (d) x^{-1} (e) x^{-2} (f) $x^{-\frac{1}{2}}$

Rules for manipulating indices

product rule $\qquad a^m \times a^n = a^{m+n}$

quotient rule

$$a^m \div a^n = \frac{a^m}{a^n} = a^{m-n}$$

power rule $\qquad (a^m)^n = a^{mn}$

Important conventions

$$\sqrt{a} = a^{\frac{1}{2}}$$

$$\sqrt[n]{a} = a^{\frac{1}{n}}$$

$$\frac{1}{x} = x^{-1}$$

$$\frac{1}{x^n} = x^{-n}$$

$$x^o = 1$$

Exercise 2A

1.　Simplify, using the three laws for indices and then check your answers with DERIVE.

(a)　$x^3 \times x^2$ 　　　　　(b)　$a^4 \times a^2 \times a^5$ 　　　　(c)　$(4a)^2 \times (2a)^3$

(d)　$6x^5 \div 2x^2$ 　　　　(e)　$c^7 \div c^3$ 　　　　　　(f)　$(3^4)^2$

(g)　$(a^2)^3$ 　　　　　　(h)　$(a^2 b^3)^4$ 　　　　　(i)　5^{-1}

(j)　2^{-4} 　　　　　　(k)　$a^{-1} \div a^{-3}$ 　　　　(l)　$\dfrac{4a^{-2} \times 2a^{-3}}{8a^{-4}}$

(m)　$(mg)^2 \div (g/m)^{-1}$ 　(n)　$a^x(a^{2x} - a^{-2x})$ 　(o)　$\left(\dfrac{x^2 y}{a^3}\right)^{-2}.$

2.   Write each of the following in the form x^n

    (a) $\sqrt{x}$ (b) $x^2\sqrt{x}$ (c) $\dfrac{x}{\sqrt{x}}$

    (d) $\left(x^3\right)^2$ (e) $\left(\sqrt{x}\right)^4$ (f) $\dfrac{x^2\sqrt{x}}{x^3}$.

3.   Evaluate the following without using a calculator

    (a) $9^{\frac{1}{2}}$ (b) $8^{\frac{1}{3}}$ (c) $16^{\frac{1}{4}}$

    (d) $100^{\frac{3}{2}}$ (e) $\dfrac{1}{(1000)^{-\frac{1}{3}}}$ (f) $4^{-\frac{1}{2}}$

    (g) $a^\circ \times b^\circ$ (h) $x^\circ + 3a^\circ + b^\circ$ (i) $\left(\dfrac{x^\circ}{a^\circ}\right)^{13}$.

    Now check your answers using DERIVE.

4.   Write each of the following numbers as decimals to four significant figures.

    (a) 3.2×10^5 (b) 1.473×10^{-4}

    (c) 9.81×10^3 (d) 1.03×10^{-6}

    (e) $(6.01 \times 10^4) \times (3.2 \times 10^6)$ (f) $(5.132 \times 10^9) \div (1.62 \times 10^3)$

    (g) $(2.43 \times 10^5)^2$ (h) $(4.72 \times 10^6) + (1.96 \times 10^4)$

5.   Write the answers to each of the following in standard form $A \times 10^n$ where $|A| < 10$ and n is an integer.

(a)   The kinetic energy of a train of mass 300 000 kg moving with speed 50 ms^{-1}. Note that kinetic energy $= \frac{1}{2}mv^2$.

(b)   The wavelength associated with electrons is given by $\lambda = h/p$ where h is Planck's constant 6.63×10^{-34} Js and p is the momentum of the electrons. Calculate the wavelength for electrons with momentum 2.1×10^{-23} Ns.

(c)   Newton's law of gravitation gives the force on the moon due to the earth as

$$F = \frac{GMm}{r^2}$$

where G is the gravitational constant 6.67×10^{-11} Nm^2 kg^{-2}, M is the mass of the earth 5.98×10^{24} kg and r is the distance of the moon from the centre of the earth 3.84×10^8 m. Calculate F.

(d)   The kinetic energy of an electron of mass 9.1×10^{-31} kg moving with speed 2×10^4 ms^{-1}.

(e)   The force on a rocket of mass 30000 kg accelerating at 21 ms^{-2}.

(Note that Newton's second law of motion gives force = mass × acceleration).

(f)   The number of seconds in one year (of 365 days).

2.2   THE EXPONENTIAL FUNCTION e^x

DERIVE ACTIVITY 2A

In this activity you will investigate exponential functions and their graphs.

Load DERIVE so that you have an algebra window and a graphics window. Use the **Options Precision** command to choose the approximate mode.

(A)   (i)   **Author** and **Plot** 2^x.
              Describe the properties of 2^x when (a) x is large and negative, (b) when $x = 0$ and (c) when x is large and positive.

       (ii)   Return to the Algebra the window and **Author** the next expression which is 3^x.

              Now **Plot** it on the same grid as 2^x.

              Repeat activity (ii) for the expressions 5^x and 10^x.

              What features do all four graphs have in common?
              How are they different?

              Clean both windows. **Delete All** the graphs and **Remove** all the expressions.

(B)   Repeat the same procedure for the following expressions

              $2^{-x}, 3^{-x}, 5^{-x}$ and 10^{-x}.

       What features do these graphs have in common?
       How are they different?
       How do they compare with the graphs drawn in (A)?

Summary

The exponential function $f(x) = a^x$ with $a > 1$ can be used to model quantities which grow larger as x increases. The shape of the graph of a^x suggests that the graph becomes more steep as x increases.

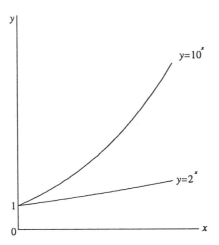

Figure 2.4   Graphs of $y = 2^x$, $y = 10^x$

For $0 < a < 1$ the graph of $f(x) = a^x$ decreases as x increases so that a^x for $a < 1$ can be used to model quantities which decay. For example, $T(t) = 0.16^t$ could model the temperature of a bowl of hot soup as a function of time, t, as it cools down.

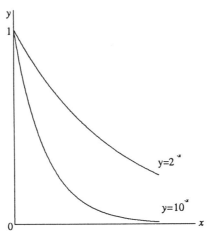

Figure 2.5   Graphs of $y = 2^{-x}$, $y = 10^{-x}$

DERIVE ACTIVITY 2B

In this activity you will investigate the gradient of the graphs of exponential functions.

In DERIVE use the **Options** and **Precision** commands to choose the **Approximate** mode. Use the **Transfer** command to **Load** the **Utility** file DIF_APPS.MTH so that you can use the TANGENT application, which gives the equation of the tangent to a curve at a chosen point.

(A) (i)   **Author** and **Plot** 2^x.

     (ii) In the Algebra window **Author** TANGENT $(2^x, x, 0)$ and **Simplify** this
           expression.
           Plot the graph of the tangent to 2^x at $x = 0$.
           Write down the slope of the tangent at $x = 0$.

     (iii) **Author** TANGENT $(2^x, x, 1)$ and **Simplify**.
           Plot the graph of the tangent to 2^x at $x = 1$.
           Write down the slope of the tangent at $x = 1$.

     (iv) Use the TANGENT application to complete Table 2.3.

Table 2.3

x	$y = 2^x$	slope of tangent
0	1	0.693147
1	2	
2	4	
3	8	

(B) (i)   Repeat activity (iv) for the function $y = 3^x$. (There is no need to draw the
           tangents at each point).
           Does your table confirm the statement in the summary that the graphs
           become more steep as x increases?

           Note from your tables that for the function 2^x the slopes are less than the
           values of the function whereas for 3^x the slopes are greater than the
           function values.

(C)   (i)   The next function is rather special.
              Delete All your graphs, return to the Algebra window and **Remove** all
              youre expressions.
              Author $\hat{e}^x$ by holding down the Alt key while you type the letter e. A
              small hat on the e will appear.
              You should see $\hat{e}^x$ in the algebra window. This is a special function in
              mathematics called *the exponential function*. DERIVE uses the notation
              $\hat{e}^x$ to denote this special function but it is normally written as e^x.
              Now **Plot** the graph of e^x.

              Compare this graph with the graphs of 2^x and 3^x.
              What can you deduce about the properties of e^x?

       (ii)   Use the TANGENT application to complete Table 2.4.

                                 Table 2.4

x	$y = e^x$	slope of tangent
0		1
1		
2		
3		

              [To find the values of e^x use the **Manage Substitute** commands and the
              approX command to give a decimal answer].

              Comment on the results in your table.

(D)   Repeat activity (C) for the special function e^{-x}. (Remember to use Alt e).

Summary

DERIVE Activity 2B introduces a very important function in mathematics denoted by e^x. It is called *the exponential function*. The number e is an irrational number (like for example π) and to 12 decimal places is given by

$$e = 2.718281828459 \ .$$

Note that DERIVE gives e to six significant figures as 2.71828.

The important property of the exponential function that makes it one of the most useful models in mathematics was shown in activity (C). This property is associated with the slope of the tangent to the graph of a function $f(x)$ called *the rate of change of $f(x)$* **with respect to x.**

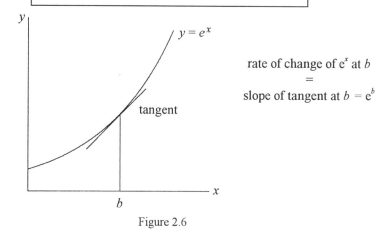

The rate of change of the function $y = e^x$ with respect to x is equal to the same function e^x.

rate of change of e^x at b
=
slope of tangent at $b = e^b$

Figure 2.6

So what this means is that the rate of change of e^x with respect to x at $x = 2$ is actually equal to e^2. Compare this with the functions 2^x and 3^x say. The rate of change of 2^x is always less than the function value whereas the rate of change of 3^x is always greater than the function value.

In theory we can model any growth or decay function by a power law a^x for some number a. This turns out to be very inconvenient and it is more usual to use the base e.

Suppose that the model for a physical quantity is

$$y = Aa^x.$$

This can be rewritten as

$$y = Ae^{kx}$$

where $a = e^k$ since $a^x = (e^k)^x = e^{kx}$.
The general shape of the graph of the functions $y = Ae^{kx}$ for $k > 0$ and $k < 0$ are shown in Figure 2.7.

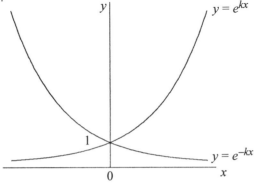

Figure 2.7

Exercise 2B

1.   Scientific calculators contain the exponential function as a special key. It is usually labelled e^x. Use your calculator to find the following as decimals to four significant figures.

(a) e^0 (b) e^2 (c) e^{-3} (d) $e^{1.6}$ (e) $e^{0.21}$

(f) $e^{0.6}$ (g) e^1 (h) $\sqrt{e}$ (i) e^4 (j) $e^{-0.2}$

Check your answers using DERIVE.

2.   Use DERIVE to draw graphs of the following functions.

   (a) e^{2x} (b) $5e^{3x}$ (c) e^{-x} (d) $4e^{-2x}$

3.   Consider the expression $f(n) = \left(1 + \dfrac{1}{n}\right)^n$. Using DERIVE complete Table 2.5
      and show that as n increases, $f(n)$ tends towards e^1.

Table 2.5

n	$\left(1+\dfrac{1}{n}\right)^n$
1	2
1.5	2.1516574
2	
3	
4	
5	
10	
100	
1000	
10000	

4.   Use the Tangent Utility in DERIVE to find the rate of change of the following
      as decimals to four significant figures.

   (a) 2^t at $t = 1.5$ (b) e^t at $t = 2.1$

   (c) e^{3x} at $x = 2$ (d) 4^{-x} at $x = 3$

5.   A function of importance in statistics is one of the form $f(x) = e^{-x^2}$. Use
      DERIVE to sketch a graph of this function.

2.3   LOGARITHMIC FUNCTIONS

Consider the exponential function $y = e^x$. If we are given a value of x, 1.5 say, then we can calculate y

$$y = e^{1.5} = 4.48168 \ .$$

However, suppose that we are given $y = 4.48168$ how do we show that $x = 1.5$? We would need to solve the equation

$$e^x = 4.48168 \ .$$

Figure 2.8 shows a diagram of the mathematical problem.

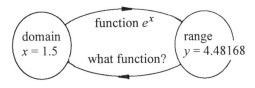

Figure 2.8

We are looking for the inverse function to e^x. For exponential functions the inverse functions are called *logarithmic functions*.

Definition of a Logarithmic Function

$$y = \log_a x \qquad a > 0, \ a \neq 1$$

is the inverse of the function $x = a^y$.

Although the index a could take any value, there are two bases which are most commonly used in mathematics,

$a = 10$ $\log_{10} x$ are called *common logarithms*

$a = e$ $\log_e x$ are called *natural logarithms* and are written as $\ln(x)$.

Hence to solve the equation $e^x = 4.48168$ we use the natural logarithm to give $x = \ln(4.48168) = 1.5$.

DERIVE ACTIVITY 2C

In this activity you will use DERIVE to investigate logarithmic functions and their graphs.

(A)   The logarithmic function to base a is written as $\log(x,a)$ and the natural logarithm is $\ln(x)$.

   (i)   **Author** and **Plot** $\log(x,2)$.
         Describe the properties of $\log_2 x$.

   (ii) **Author** and **Plot** $\log(x,3)$.

   (iii) Repeat for the expressions $\log(x,10)$ and $\log(x,\hat{e})$ (remember to input e as Alt e).

   Compare the four graphs. What features are the same and what features are different?

   For what value of x do the graphs cut the x-axis? Use the definition of logarithmic functions to explain your answer.

(B)   **Delete All** your graphs and **Remove** your expressions.
      Author and **Plot** $\log(x,e)$ and $\ln(x)$ on the same grid.

   What happens? What does this mean about the two functions?

(C)   (i)   **Delete All** your graphs and **Remove** your expressions.
            Author and **Plot** e^x and $\ln(x)$.
            Compare the graphs of e^x and $\ln(x)$.
            Author and **Plot** x.
            Note that the graph of $y = \ln(x)$ is a reflection of the graph of $y = e^x$ in the line $y = x$.

   (ii) Repeat activity (i) for other exponential functions and their associated logarithmic functions.
         Do you see the same reflection property?

(D)   **Delete All** your graphs and **Remove** your expressions.
      Author and **Plot** $y = \log_2 x$ and $y = 1$.
      Move the cross in window 2 to find the value of x where the graphs cross.
      Repeat for $y = \log_3(x)$ and $y = \log_e(x)$.
      Deduce a general result for the solution of $\log_a(x) = 1$.

Summary

1. The graph of each logarithmic function has the general shape shown in Figure 2.9.

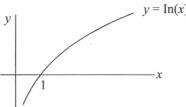

Figure 2.9   The graph of $y = \ln(x)$

2. For any base $\log_a(1) = 0$ and $\log_a(a) = 1$.

Exercise 2C

1. Common and natural logarithms are available on your calculator. They are usually labelled log and ln respectively.

    Use your calculator to find the values of the following logarithms.

    (a) $\log(2)$ (b) $\log(10)$ (c) $\log(4.2)$ (d) $\log(0.7)$

    (e) $\ln(2)$ (f) $\ln(4.2)$ (g) $\ln(0.7)$ (h) $\ln(1)$

    (i) $\log(1000)$ (j) $\ln(e^2)$ (k) $\log(1)$ (l) $\ln(e)$

2. Use the definition of logarithmic functions to show that $\log_a(a) = 1$ for any base a. Check the result using DERIVE and several values for a.

3. Solve the following equations.

    (a) $3.1 = 10^x$ (b) $0.2 = 10^x$

    (c) $1.7 = e^x$ (d) $5.1 = e^x$

    (e) $11.2 = e^t$ (f) $0.47 = e^t$

    (g) $\ln(t) = 1.7$ (h) $\ln(x) = 4.2$

    (i) $\log(3x) = 6$ (j) $\ln(5t) = 3.1$

4.   The charge on a capacitor as it is charged by a battery is given by

$$Q = Q_0(1 - e^{-50t})$$

where Q_0 is a constant.
Interpret the quantity Q_0.
Calculate the time when $Q/Q_0 = 0.5$.
Use DERIVE to **Plot** a graph of charge against time when $Q_0 = 10$.
Use the **Scale** and **Centre** commands to focus on the relevant part of the graph.
What happens to Q as t becomes large?

5.   Radioactive decay of substances is modelled by the exponential function. For a
sample of iron nuclide the fraction remaining as a function of time is modelled
by

$$N = e^{-0.25t} .$$

Calculate the time when $N = 0.5$. This is called the *half-life* of the decay.

6.   Radioactive carbon-14 decays at a rate of $1.238 \times 10^{-4} \text{yr}^{-1}$ so that the fraction of
carbon-14 in a sample is modelled by

$$M = e^{-1.238 \times 10^{-4} t} .$$

Calculate the half life of carbon-14.
Calculate the times for the fraction of carbon-14 in the sample to reduce to 10%
and 5%.

7.   The atmospheric pressure at a height h (in km) above sea level obeys the rule

$$p = p_0 e^{-0.15h}$$

where p_0 is the pressure at sea level, $p_0 = 100\,000$ pascals.

(a)   Calculate the pressure at heights 1 km, 2 km, 5 km and 10 km.

(b)   Draw a graph of p against h.

(c)   At what height is the pressure equal to half the value at sea level?

(d)   At what height is the pressure equal to one-tenth of the value at sea
level?

2.4   RULES FOR LOGARITHMS

The following two properties of logarithms have already been observed.

$\ln(1) = 0$	$\log_{10}(1) = 0$
$\ln(e) = 1$	$\log_{10}(10) = 1$

MINI INVESTIGATION

(A)   Before starting use the **Manage** and **Logarithm** commands. Select **Collect** from the menu you obtain.

    (i)   Using DERIVE, **Author** and **Simplify** the following expressions.

        (a) $\ln(2) + \ln(3)$ (b) $\ln(5) + \ln(10)$

        (c) $\ln(6) + \ln(2)$ (d) $\ln(2) + \ln(x)$

        What do you observe?

    (ii)   **Author** and **Simplify**

        (a) $\ln(4) - \ln(2)$ (b) $\ln(8) - \ln(2)$

        (c) $\ln(1000) - \ln(10)$ (d) $\ln(x) - \ln(3)$

        What do you observe?

    (iii)   **Author** and **Simplify**

        (a) $3\ln(4)$ (b) $4\ln(2)$

        (c) $2\ln(9)$ (d) $\frac{1}{2}\ln(64)$

        (e) $3\ln(1)$ (f) $\frac{1}{5}\ln(32)$

        What do you observe?

(B)   Use the **Manage** and **Logarithm** commands. Select **Expand** from the menu

    Author and **Expand**

    (a) $\ln(1)$ (b) $\ln(9)$ (c) $\ln(2.5)$

    (d) $\ln(\sqrt{2})$ (e) $\ln(12)$ (f) $\ln(24)$

    (g) $\ln(2x)$ (h) $\ln\left(\dfrac{x}{5}\right)$ (i) $\ln\left(\dfrac{3x}{2}\right)$

(C)   Do the results change if logarithms to a different base are used?

Summary

    You will have observed that there are three important rules for manipulating logarithms. These rules are:

product rule	$\log_a(xy) = \log_a x + \log_a y$
quotient rule	$\log_a(x/y) = \log_a x - \log_a y$
power rule	$\log_a(x^r) = r\log_a x$

To prove these three rules we use the rules for manipulating indices.

Let $p = \log_a(x)$ and $q = \log_a(y)$, then using the definition of logarithms $x = a^p$ and $y = a^q$.

Product rule   Multiplying x and y we have

$$xy = a^p a^q = a^{p+q}.$$

    Then

$$\log_a(xy) = p + q$$
$$= \log_a(x) + \log_a(y).$$

Quotient rule   Dividing x by y we have

$$\frac{x}{y} = \frac{a^p}{a^q} = a^{p-q} \ .$$

Then

$$\log_a (x/y) = p - q$$
$$= \log_a (x) - \log_a (y) \ .$$

Power rule $x^r = \left(a^p\right)^r = a^{pr}$

Then

$$\log_a x^r = pr = r\log_a (x) \ .$$

Exercise 2D

1.   Expand the following expressions using the rules of logarithms.

    (a) $\ln(a^3 b^2)$ (b) $\ln(0.1x^2)$

    (c) $\ln(3.7t/a)$ (d) $\ln(3p(x+y)^2)$

    (e) $\log(\sqrt{2}s/t)$ (f) $\log(0.2v^2)$

    (g) $\log(10xy^2)$ (h) $\ln(\frac{1}{2}at^2)$

    (i) $\log(3(a+b)/s)$

2.   Write each of the following as a single logarithm and simplify.

    (a) $\log(40) - \log(5)$ (b) $\log(10x) + \log(2x) - \log(x)$

    (c) $\ln(5a) + \ln(2b) - \ln(c)$ (d) $3\ln(a) + 5\ln(b)$

    (e) $2\ln(x) - 4\ln(y)$ (f) $\log(x) - \log(y) + 1.5\log(a)$

3.   Solve the following equations.

   (a) $\ln(2x) = 5.6$

   (b) $2\ln(3x) - 3\ln(x) = 1.5$

   (c) $3\log(x) + 2\log(3x) = 3.1$

2.5   MODELLING WITH POWER AND EXPONENTIAL LAWS

Scientific laws usually arise in one of two ways. Either existing scientific theories are used to develop new theories which are tested by experiment or results of experiments are used directly to formulate empirical models. In both of these approaches the use of logarithms can play an important part.

In Chapter 1 we saw that if data for two variables fits a straight line then finding its equation and hence the relation between the variables is $y = mx + c$ where m is the slope of the line and c is the intercept. But experimental data does not always give straight lines. However, in situations where the model is a power law $y = ax^b$ or an exponential law $y = ae^{bx}$ the use of logarithms will transform the graphs to straight lines.

Power Law. Suppose that the relationship between two variables is thought to be a power law of the form $y = ax^b$. Then a graph of y against x will be of the form shown in Figure 2.10. An important clue to a power law is that the graph passes through the origin.

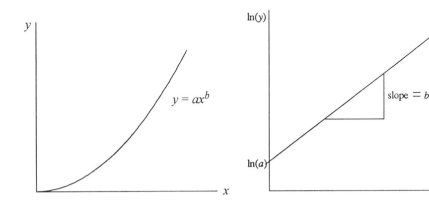

Figure 2.10   Typical power law graph   Figure 2.11   For a power law $y = ax^b$ a graph
                $y = ax^b$ of $\ln(y)$ against $\ln(x)$ will be linear

If we take (natural) logarithms of each side of the equation we get

$$\ln(y) = \ln(ax^b) = \ln(a) + \ln(x^b)$$
$$= \ln(a) + b\ln(x).$$

Thus a graph of $\ln(y)$ against $\ln(x)$ will be a straight line (see figure 2.11).
        The slope of the straight line graph is the index b and the intercept $\ln(a)$ can be used to find a.

Exponential Law. Suppose that the relationship between the two variables x and y is thought to be exponential of the form $y = ae^{bx}$. Then a graph of y against x will be of the form shown in Figure 2.12.

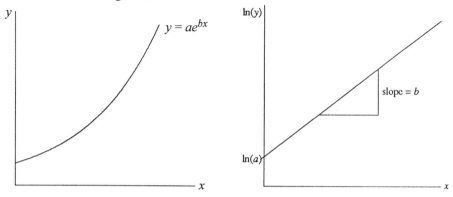

Figure 2.12   Typical exponential law   Figure 2.13   For an exponential law $y = ae^{bx}$ a
                    graph $y = ae^{bx}$ graph of $\ln(y)$ against x will be linear

If we take (natural) logarithms of each side we get

$$\ln(y) = \ln(a\,e^{bx}) = \ln(a) + \ln(e^{bx})$$
$$= \ln(a) + bx.$$

Thus a graph of $\ln(y)$ against x will be a straight line. The slope and intercept of the straight line graph can be used to find a and b.
        The following example shows how we use DERIVE to find relationships between variables.

Example 2A

Table 2.6 shows three sets of experimental data, one pair satisfies a power law, one pair satisfies an exponential law and the other pair does not satisfy either of these laws.

Use appropriate graphs to find the relationships.

Table 2.6

x	0.1	0.4	0.7	1.0	1.3	1.6
y	2.010	0.606	0.182	0.055	0.017	0.005

u	0.1	0.3	0.6	0.9	1.2	1.5
v	1.00	0.96	0.83	0.62	0.36	0.07

t	0.1	0.5	0.9	1.3	1.7	2.1
s	0.084	1.293	3.511	6.561	10.35	14.83

Solution

The first step is to enter the data in DERIVE as three matrices and to plot the graphs of the data. (Use **Declare Matrix** responding with 6 rows and 2 columns). Figure 2.14 shows a screen printout.

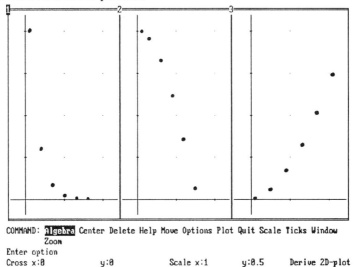

```
COMMAND: Algebra Center Delete Help Move Options Plot Quit Scale Ticks Window
         Zoom
Enter option
Cross x:0            y:0            Scale x:1       y:0.5      Derive 2D-plot
```

Figure 2.14   DERIVE graphs of the data

The graph in window 3 suggests that s and t are related by a power law $s = at^b$ and the graph in window 1 suggest that x and y are related by an exponential law $y = ae^{bx}$. The relation between u and v, shown in window 2, is neither of these.

Consider the x,y data. Since we expect an exponential law $y = ae^{bx}$ we use DERIVE's Fit command to find a linear law between $\ln(y)$ and x. Figure 2.15 shows the screen for this activity.

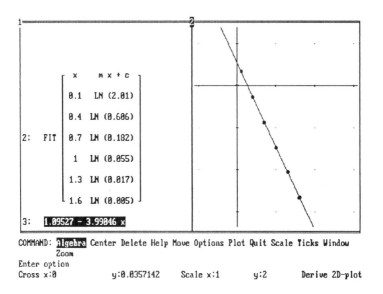

Figure 2.15

We have

$$\ln(y) = 1.09527 - 3.99046x .$$

Comparing with the general form

$$\ln(y) = \ln(a) + bx .$$

We deduce that $b = -3.99$ and $\ln(a) = 1.095$ ie. $a = e^{1.095} = 2.99$ giving the relationship

$$y = 2.99e^{-3.99x} .$$

[It is likely that the law $y = 3e^{-4x}$ would be a very good model for this data].

Consider the t,s data. Since we expect a power law $s = at^b$ we use DERIVE's Fit command to find a linear law between $\ln(s)$ and $\ln(t)$.

Figure 2.16 shows the screen for this activity. (In this printout $x = \ln(t)$ and $y = \ln(s)$).

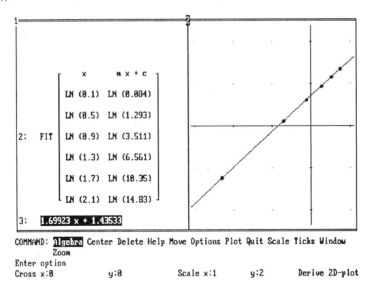

Figure 2.16

We have

$$\ln(s) = 1.69923\ln(t) + 1.43533 .$$

Comparing this with the general form

$$\ln(s) = \ln(a) + b\ln(t) .$$

We deduce that $b = 1.699$ and $\ln(a) = 1.435$ so that $a = e^{1.435} = 4.2$ giving the relationship

$$s = 4.2t^{1.699} .$$

[It is likely that the law $s = 4.2t^{1.7}$ would be a very good model for this data].

Exercise 2E

1. Which of the curves in Figure 2.17 suggests (a) a linear law, (b) a power law, (c) an exponential law, (d) none of these.

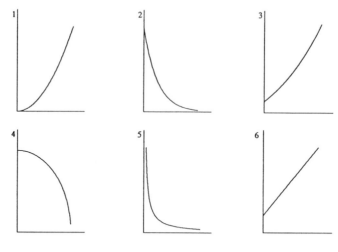

Figure 2.17

2. Find the power law formula between the variables given in Table 2.7.

Table 2.7

(a)	x	1	2	3	4	5	6
	y	3.42	12.76	27.58	47.64	72.79	102.9

(b)	s	0.3	1.1	1.9	2.3	3.2	4.1
	t	16.7	1.24	0.42	0.28	0.15	0.089

3. Find the exponential law formula between the variables given in Table 2.8.

Table 2.8

(a)	u	2	4	6	8	10
	v	2.68	1.80	1.20	0.808	0.541

(b)	x	0.1	0.3	0.7	0.9	1.6
	y	7.67	9.75	15.75	20.02	46.38

4.   The data in Table 2.9 shows the distance from the Sun and the period for each of the planets in the solar system.

Table 2.9

Planet	Distance, R (millions of km)	Period, T (days)
Mercury	57.9	88
Venus	108.2	225
Earth	149.6	365
Mars	227.9	687
Jupiter	778.3	4329
Saturn	1427	10753
Uranus	2870	30660
Neptune	4497	60150
Pluto	5907	90470

Find a relationship between T and R.

5.   Table 2.10 contains data that was produced in an experiment where the saturated vapour pressure of a fixed volume of water was determined at different temperatures.

Table 2.10

S.V.P. k (Nm^{-2})	0.61	0.86	1.21	1.70	2.33
Temperature T $(°C)$	0	5	10	15	20

Find a possible scientific model relating S.V.P. and temperature for this constant volume of water.

6.   Table 2.11 shows experimental values for the potential difference V and
        current I for a semiconductor diode.

Table 2.11

potential difference V (volts)	current I (microamps)
0.255	0.40
0.315	1.60
0.345	3.6
0.385	8.9
0.410	18.2
0.455	52.2
0.475	90.3
0.495	140
0.505	182
0.515	223
0.530	310

The exponential model $I = I_0 e^{aV}$ is proposed for this data. Show that this is a
good model and find values of I_0 and a.

7.   The data in Table 2.12 gives the atmospheric pressure, expressed as a
        percentage of its sea-level value, at various altitudes.

Table 2.12

altitude, h (km)	0	5	10	14	20	24	30
pressure, p (% of sea level value)	100	53.0	26.0	14.0	5.4	2.9	1.2

Find a possible scientific model relating pressure and altitude.

8.   An experiment on the decomposition of nitrous oxide yielded the following
        values for the velocity constant k at various temperatures T.

Table 2.13

T(K)	985	1005	1058	1069	1105
k(mol min^{-1})	0.224	0.447	2.00	2.52	6.31

For most reactions a model between k and T is $k = ae^{-b/T}$ where a and b are
constants. Show that this data fits such a model and find the values of a and b.

3

Trigonometric functions

3.1   INTRODUCTION

Oscillations occur frequently in real life, for example the swinging of a pendulum in a clock, the changing heights of tides, the movement of a needle in a sewing machine, and the amount of daylight as the seasons change. We will see that these oscillations can be modelled by the familiar sine and cosine functions.

3.2   DEGREES AND RADIANS

For most practical purposes angles are measured in degrees. Indeed, this has been the custom for about 4000 years, since the time of the Babylonian civilisation.

However, for the purposes of advanced work in mathematics another unit is needed. It is called the *radian*.

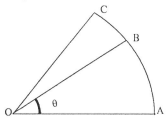

Figure 3.1

In figure 3.1 the relative sizes of angles AOB and BOC may be compared by measuring the arc lengths AB and BC instead of counting the number of degrees turned through in each angle, provided $OA = OB = OC$. This is the essence of radian measure. Radians are arc lengths. In figure 3.1 if arc length AB is equal to radial length OA then angle AOB is 1 radian. If the length of the arc AC is $2OA$ then the angle turned through is 2 radians, and so on.

In general the angle θ is given in radians by

$$\theta = \frac{\text{arc length } AB}{\text{radius } OA} \; .$$

To find a connection between degrees and radians consider a circle of radius r.

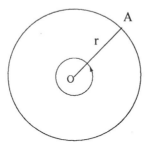

Figure 3.2

In one rotation of the radial line OA around the circle an angle of $360°$ is turned through. Since the radius of the circle is r, the distance travelled around the circle in one revolution is $2\pi r$. Then

$$\text{angle turned through} = \frac{\text{circumference}}{\text{radius } OA}$$

$$360° = \frac{2\pi r}{r} \; .$$

Thus 2π radians $= 360°$, or π radians $= 180°$, or 1 radian $= 180/\pi$ degrees $(= 57.3°)$.

So, if angle AOB in figure 3.3 is measured in degrees then

$$\theta \text{ degrees} = \frac{\pi\theta}{180} \text{ radians.}$$

On the other hand, if angle AOB is measured in radians then

$$\theta \text{ radians} = \frac{180\,\theta}{\pi} \text{ degrees.}$$

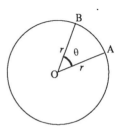

Figure 3.3

It is very common in advanced mathematics to state radians as multiples of π. For example,

$$90° = \frac{\pi}{180} \times 90 \text{ radians}$$

$$= \frac{\pi}{2} \text{ radians}$$

It is important when working with angles on your calculator to ensure that your calculator knows exactly which units you are using. There is often a key (or switch) to define degrees (often written DEG) or radians (often written RAD).

Consider again Figure 3.3; from the definitions of radians

$$\theta = \frac{\text{arc length } AB}{\text{radius}} \ .$$

If the radius of the circle is r then we have the important result

---
arc length $AB = r\theta$

where θ is measured in radians.

The area of the circle is πr^2 and the sector OAB is a fraction of the circle given by

$$\frac{\text{area sector } OAB}{\text{area of circle}} = \frac{\theta}{2\pi} \ .$$

Hence the area of the sector OAB is given by

---
area sector $OAB = \frac{1}{2}r^2\theta$

where θ is measured in radians.

Exercise 3A

1. Use the conversion formulae given above and your calculator to change degrees into radians or radians into degrees in this problem.

    (a) 65°   (b) 18°   (c) 1.5 radians   (d) 0.5 radians   (e) 2.8 radians

    (f) 150°   (g) 5 radians   (h) 279°   (i) 1 radian   (j) 1°

2. Complete the following table of degree and radian equivalence.

    Table 3.1

Degrees	0	30	45	60	90	120	135	150	180
Radians	0		$\dfrac{\pi}{4}$		$\dfrac{\pi}{2}$				π

Degrees	210	225	240	270	300	315	330	360
Radians		$\dfrac{5\pi}{4}$			$\dfrac{5\pi}{3}$		2π	

3. Find in radians, the angle subtended at the centre of a circle of radius 20 cm by an arc:

    (a) 8 cm long   (b) 35 cm long.

4. Find in radians the angle of a sector of arc length 8 cm in a circle of radius 12 cm. What is the area of the sector?

5. The area of a sector of a circle of radius 15 cm is 9 cm^2. Calculate length of the arc of the sector.

3.3   RIGHT ANGLED TRIANGLES

In elementary mathematics you will be familiar with the trigonometric ratios *sine, cosine* and *tangent* which are defined in terms of the sides of a right-angled triangle.

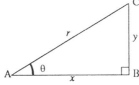

Figure 3.4

Thus for the right-angled triangle in figure 3.4 we have

$$\sin\theta = \frac{y}{r} \qquad \cos\theta = \frac{x}{r} \qquad \tan\theta = \frac{y}{x}.$$

You will also be familiar with Pythagoras' theorem $x^2 + y^2 = r^2$.

DERIVE ACTIVITY 3A

DERIVE will always assume that you are working in radians when using trigonometry. The first parts of this activity show you how to use degrees and the later parts deal with simple problems involving right angled triangles.

(A)   **Author** 44deg. **Simplify** this expression to obtain $\dfrac{11\pi}{45}$. This is the way DERIVE will convert degrees into radians. Applying **approX** will produce a decimal approximation of the radian value.

(B)   **Author** 1.5/deg. Use **approX** on this expression to obtain 85.9436. In this way DERIVE converts radians into degrees.

Use DERIVE to convert degrees into radians or radians into degrees in this exercise. N.B. To obtain π press Alt *P*.

(i)   82°   (ii)   3.72 radians   (iii) $\dfrac{7\pi}{5}$ radians

(iv)   250°   (v)   310°   (vi)   0.263 radians

(vii) 1 radian   (viii) 1°   (ix)   10.6°

(x) $\dfrac{8\pi}{15}$ radians

(C)   Figure 3.5 shows a ladder 3.5 m long leaning against a wall and making an
       angle of 61.3° with the ground. How far out from the wall is the foot of the
       ladder and how far up the wall is the top of the ladder?

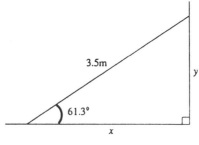

Figure 3.5

The length marked x denotes the distance of the foot of the ladder from the wall
and the length marked y represents the height of the top of the ladder up the
wall. To find x we recognise that the cosine function must be used.

Author $\cos\theta = \dfrac{x}{r}$. (To obtain θ press Alt H). **Manage Substitute** 3.5 for r
and 61.3 deg for θ. **soLve** for x and **approX**. The value $x = 1.68$ (to 3 sf) is
obtained.

To find y we must use sine. **Author** $\sin\theta = \dfrac{y}{r}$, **Manage Substitute** the
appropriate values, **soLve** and **approX** to obtain the value $y = 3.07$ (to 3 sf).

Thus the ladder is 1.68 m out from the wall and reaches 3.07 m up the wall.

(D)   A rigid strut of length r metres is to support a bench 0.75 m wide. The strut is
       fastened to the end of the bench and 0.5 m below its contact with the wall, as
       shown in figure 3.6. How long is the strut and what angle θ does it make with
       the wall?

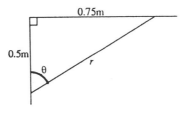

Figure 3.6

To find r we need to use Pythagoras' Theorem.

Author $x^2 + y^2 = r^2$. **Manage Substitute** 0.5 for x and 0.75 for y. **soLve** for r ignoring the negative result. Using **approX** gives the radius $r = 0.901$ (to 3 sf).

Author $\tan\theta = \dfrac{y}{x}$. **Manage Substitute** the values for x and y. **soLve** for θ.

Three values in radians are returned (if in exact precision mode). Use **approX** on each one of them and convert the answers to degrees. It can be seen that the only possible value for θ in this situation is $\theta = 56.3$.

Thus the strut needs to be 0.901m long and inclined at $56.3°$ to the wall.

Now solve the following problems.

(E)   A rectangular frame measures 9 m by 5 m and is kept in shape by two diagonal struts.

    (i)   Find the length of a strut.

    (ii)   What is the angle made by a strut with the longest side of the rectangle?

(F)   The road from a village P runs due east for 3 miles to a fort Q. There is a television mast due north of Q and 6 miles from Q. Find the distance and bearing of the mast from P.

(G)   A river runs parallel to a software plant. The roof of this plant is 35 m above the water level of the river. From a point on the roof, looking directly across the river, the angle of depression of the nearer bank is $39°$ and the angle of depression of the farther bank is $19°$. Find the width of the river.

(H)   In cartesian coordinates L is the point $(-2,-3)$, M is the point $(1,1)$ and N is the point $(2,-1)$. Find

    (i)   the angle between LM and a line parallel to the y axis,

    (ii)   the angle between MN and a line parallel to the x axis,

    (iii)   hence find angle LMN.

3.4   NON RIGHT ANGLED TRIANGLES

It is obvious that not all triangles are right angled and so it becomes necessary to develop further trigonometric techniques to deal with any sort of triangle. Two important laws can be used in problems which do not involve right angled triangles; these are *the sine rule* and *the cosine rule*.

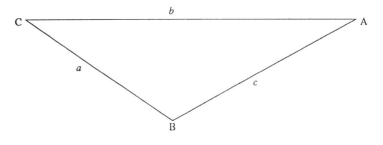

Figure 3.7

Figure 3.7 shows a helpful convention when dealing with non right angled triangles. The sides opposite the vertices A, B and C are designated a, b and c. Such a triangle is called *an oblique triangle*.

In figure 3.8 the point D is the foot of the perpendicular line drawn from B to AC. The length of BD is h. If the length of AD is x then the length of CD will be $b - x$.

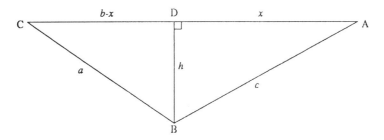

Figure 3.8

From triangles ABD and CBD,

$$\sin A = \frac{h}{c} \quad \text{and} \quad \sin C = \frac{h}{a}$$

so $c \sin A = h$ and $a \sin C = h$.

Then $c \sin A = a \sin C$

or $\dfrac{c}{\sin C} = \dfrac{a}{\sin A}.$

By choosing other altitudes and following the same argument we find that

$$\frac{b}{\sin B} = \frac{a}{\sin A} \quad \text{and} \quad \frac{c}{\sin C} = \frac{b}{\sin B}.$$

These three results may be drawn together into the *Sine Rule*.

$$\boxed{\frac{a}{\sin A} = \frac{b}{\sin B} = \frac{c}{\sin C}}$$

Now we develop the cosine rule. Consider figure 3.8. Applying Pythagoras' Theorem to triangle ABD gives

$$h^2 = c^2 - x^2$$

and to triangle CBD,

$$h^2 = a^2 - (b-x)^2$$

hence

$$a^2 - (b-x)^2 = c^2 - x^2$$

$$a^2 - b^2 + 2bx - x^2 = c^2 - x^2$$

$$a^2 = b^2 + c^2 - 2bx.$$

But from triangle ABD

$$\cos A = \frac{x}{c}$$

so

$$c \cos A = x$$

and therefore

$$a^2 = b^2 + c^2 - 2bc \cos A .$$

In the same way, using other altitudes, it can be shown that

$$b^2 = a^2 + c^2 - 2ac \cos B$$

and $$c^2 = a^2 + b^2 - 2ab \cos C .$$

These last two results are of a similar form and lead to the *Cosine Rule*.

$$a^2 = b^2 + c^2 - 2bc \cos A$$
$$b^2 = a^2 + c^2 - 2ac \cos B$$
$$c^2 = a^2 + b^2 - 2ab \cos C$$

Furthermore, the area of triangle ABC is given by

$$area = \tfrac{1}{2}bh$$

but since $h = a \sin C$ we find

$$area = \tfrac{1}{2}ab \sin C .$$

Since any altitude can be chosen, we find three versions of the area formula.

$$Area = \tfrac{1}{2}ab \sin C$$
$$= \tfrac{1}{2}bc \sin A$$
$$= \tfrac{1}{2}ac \sin B$$

Example 3A

Figure 3.9 shows a triangular field in which side AB is 400 m long, the angle at A is 65° and the angle at C is 100°.

(a)   What is the size of the angle at B?

(b)   How long are the sides BC and AC?

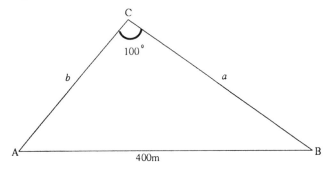

Figure 3.9

Solution

(a)   Since the angles in a triangle add up to 180°, the angle at B is 15°.

(b)   We use the sine rule in this problem to find a and b.

$$\frac{a}{\sin 65°} = \frac{b}{\sin 15°} = \frac{400}{\sin 100°}$$

Solving for a,

$$a = \frac{400 \sin 65°}{\sin 100°} = 368.115 \text{ m}.$$

Solving for b,

$$b = \frac{400 \sin 15°}{\sin 100°} = 105.124 \text{ m}$$

The lengths of the two sides are 368 m and 105 m to 3 sf.

Example 3B

Figure 3.10 shows an oblique triangle in which angle B is $135°$ and sides a and c measure 3.15 cm and 2.72 cm respectively. Find the unknown angles and side length.

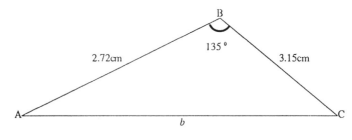

Figure 3.10

Solution

We use the cosine rule to find side length b.

$$b^2 = a^2 + c^2 - 2ac\cos B$$

$$b^2 = 3.15^2 + 2.72^2 - 2 \times 3.15 \times 2.72 \cos 135°$$
$$= 29.44 \ .$$

Hence

$$b = \sqrt{29.44} = 5.43 \text{ cm}.$$

To find angle A (or C) we use the sine rule

$$\frac{a}{\sin A} = \frac{b}{\sin B} \ .$$

Thus

$$\sin A = \frac{a \sin B}{b} = \frac{3.15 \sin 135°}{5.43} = 0.41 \ .$$

Solving for A we have

$$A = 24.2°$$

since the angles of a triangle add to $180°$, the angle at C is $20.8°$.

DERIVE ACTIVITY 3B

The aim of this activity is to show how DERIVE can be used to do calculations like those in Examples 3A and 3B. First use **Option Input** and choose **Case**: **Sensitive** so that DERIVE can recognise c and C.

(A)   Consider again Example 3A. To solve for a,

Author $a/\mathrm{SIN}A = c/\mathrm{SIN}C$.
Manage Substitute $C = 100$ deg, $A = 65$ deg, $c = 400$.
soLve to show that $a = 368.115$.

1: $\dfrac{a}{\mathrm{SIN\ (A)}} = \dfrac{c}{\mathrm{SIN\ (C)}}$

2: $\dfrac{a}{\mathrm{SIN\ (65\ °)}} = \dfrac{400}{\mathrm{SIN\ (100\ °)}}$

3:   a = 368.115

COMMAND: **Author** Build Calculus Declare Expand Factor Help Jump soLve Manage
              Options Plot Quit Remove Simplify Transfer moVe Window approX
Compute time: 0.0 seconds
Solve(2)   Free:100%   Derive Algebra

Figure 3.11

To solve for b,

Author $b/\mathrm{SIN}B = c/\mathrm{SIN}C$.
Manage Substitute $C = 100$ deg, $B = 15$ deg, $c = 400$.
soLve to show that $b = 105.124$.

(B)   Consider again Example 3B. To solve for b,

Author $b^2 = a^2 + c^2 - 2ac\ \mathrm{COS}B$.
Manage Substitute $B = 135$ deg, $a = 3.15$, $c = 2.72$.
approX to show that $b = 5.42566$ (ignore negative answer).

To find angle A use similar steps as in (A) to show that $A = 0.423033$ (in radians). This is $24.238°$.

Now use the sine rule, cosine rule and DERIVE to solve the following problems.

(C)   In triangle ABC, $A = 34.6°$, $C = 80.1°$ and $c = 10$ cm. Find a and b.

(D)   In triangle ABC, $a = 6.8$ cm, $b = 10.5$ cm and $B = 76.8°$. Find A and C and the area of the triangle.

(E)   A boat is sailing directly towards the foot of a cliff. The angle of elevation of a point on the top of the cliff, and directly ahead of the boat, increases from 6° to 10° as the boat sails 250 m. Find

    (i)   the original distance of the boat to the point on the top of the cliff,

    (ii)   the height of the cliff.

(F)   The side of a triangular field PQ is 150 m long. From P the bearing of the corner Q is 035° and the bearing of the corner R is 115°. From Q the bearing of R is 171°. Find

    (i)   the distances PR and QR,

    (ii)   the area of the field in m^2.

(G)   Find the angle between the hands of a clock at 10:20. If the hour hand is 15 cm long and the minute hand is 20 cm long, how far apart are the tips of the hands at this time?

(H)   A ship leaves port A and sails 12 nautical miles on a bearing of 025° followed by 15 nautical miles on a bearing of 192° to arrive at the next port B. Find the distance and bearing of A from B.

3.5   THE COSINE AND SINE FUNCTIONS

Trigonometry is a very powerful branch of mathematics. This becomes most evident once it is seen to break free of its apparent confinement to triangles. Anything that *oscillates*, for example a pendulum swinging or the variation of daylight with the changing seasons, may be *modelled* mathematically by means of the *cosine* and *sine* functions.

Consider a circle of unit radius called *the unit circle* with its centre at the origin of a Cartesian coordinate system (figure 3.12).

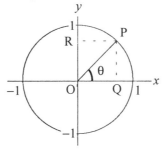

Figure 3.12

A point P on the circumference of this circle makes an angle θ with the positive direction of the x axis. For simplicity, P is shown lying in the first quadrant.

Right angled triangles OPQ and OPR are formed with the coordinate axes and since $OP = 1$, the lengths of OQ and OR are given by

$$OQ = \cos \theta \quad \text{and} \quad OR = \sin \theta .$$

If we rotate the point P in an anticlockwise direction, so that θ increases, then we can define

$$OQ = x \text{ coordinate of the point } P = \cos \theta$$
$$OR = y \text{ coordinate of the point } P = \sin \theta$$

With these definitions the values of sine and cosine can be found for any angle. The graphs of the x and y coordinates of P are shown in Figures 3.13(a) and (b). (Note that a clockwise rotation of P corresponds to negative values of θ).

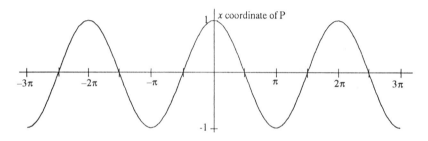

Figure 3.13(a)   Graph of x coordinate of $P = OQ$

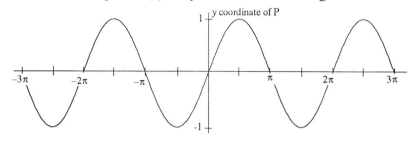

Figure 3.13(b)   Graph of y coordinate of $P = OR$

Figure 3.13(a) shows the graph of **the cosine function** for angles of any magnitude while figure 3.13(b) shows the graph of **the sine function**. From these graphs we can deduce important features of the sine and cosine functions.

1.   The graphs of both functions repeat themselves every 2π radians: they are *periodic* with a *period* of 2π

$$\cos(\theta + 2\pi) = \cos\theta \qquad \sin(\theta + 2\pi) = \sin\theta.$$

2.   The graph of the cosine function has line symmetry about its vertical axis: it is an example of an *even function*

$$\cos(-\theta) = \cos\theta .$$

3.   The graph of the sine function has a half-turn rotational symmetry about the origin: it is an example of an *odd function*

$$\sin(-\theta) = -\sin\theta .$$

4. Both graphs have a range that is limited: they are *bounded* between +1 and –1 inclusive

$$-1 \le \cos\theta \le 1 \qquad -1 \le \sin\theta \le 1.$$

5. Both graphs have no breaks in them and their domains are unlimited: they are *continuous*.

6. If the graph of the cosine function is translated to the right by $\dfrac{\pi}{2}$ radians it becomes the graph of the sine function

$$\cos\left(\theta - \frac{\pi}{2}\right) = \sin\theta \ .$$

7. If the graph of the sine function is translated to the left by $\dfrac{\pi}{2}$ radians it becomes the graph of the cosine function

$$\sin\left(\theta + \frac{\pi}{2}\right) = \cos\theta \ .$$

Exercise 3B

1. Use DERIVE to **Plot** the graphs of the sine and cosine curves for the range of values -2π to $+4\pi$. Check properties 1 to 7 above using your graphs.

2. Plot the graphs of the following functions.

(a) $y = x$ (b) $y = x^2$ (c) $y = x^3$

(d) $y = e^x$ (e) $y = e^{-x}$ (f) $y = 2$

(g) $y = x\sin x$ (h) $y = x\cos x$ (i) $y = x^2\cos x$

(j) $y = \dfrac{1}{1+x^2}$ (k) $y = e^{-x^2}$

Classify each function as even, odd, or neither even nor odd.

3.6   THE TANGENT FUNCTION

From elementary trigonometry you will know that

$$\tan\theta = \frac{\sin\theta}{\cos\theta}.$$

Now we can extend the graph of tanθ beyond the domain $0 \le \theta \le \frac{\pi}{2}$ with which you are familiar. Figure 3.14 shows the graph of tanθ for the domain $-3\pi \le \theta \le 3\pi$.

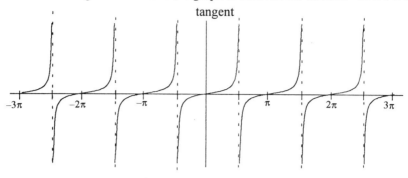

Figure 3.14   Graph of tan(θ)

Exercise 3C

1.   Use Figure 3.14 to deduce the properties of the tangent function (compare properties 1 to 5 for the sine and cosine functions).

2.   Although tanθ is not defined for θ = 90° it is defined for values of θ close to 90°.

Find the values of

(a)   tan 89°   (b)   tan 89.9°   (c)   tan 89.99°
(d)   tan 90.01°   (e)   tan 90.1°   (f)   tan 91° .

3.   At θ = 90° the graph of tanθ has an asymptote. Where else does tanθ have asymptotes?

3.7   FEATURES OF TRIGONOMETRIC FUNCTIONS

DERIVE ACTIVITY 3C

The aim of this activity is to explore the properties of the trigonometric functions and their graphs.

(A)   (i)   **Author** and **Simplify** VECTOR($n$sin$x$,n,1,3).
This will create the vector list $[\sin x, 2\sin x, 3\sin x]$.

Plot the vector with the **Scale** $x:\dfrac{\pi}{2}, y:3$.
How do the three graphs compare?
Delete the graphs and remove the expressions.

(ii)   Repeat with VECTOR($n$cos$x$,n,1,3).

(iii)   Repeat with VECTOR($n$tan$x$,n,1,3).

(B)   (i)   **Author** and **Simplify** VECTOR(sin(nx),n,1,3).

Plot the vector with the scales $x:\dfrac{\pi}{2}, y:1$. (**Plot Under** will give the best effect).
How do the three graphs compare?
Delete the graphs and remove the expressions.

(ii)   Repeat with VECTOR(cos(nx),n,1,3).

(iii)   Repeat with VECTOR(tan(nx),n,1,3).

(iv)   Try sketching by hand a graph of the function $3\sin 2x$. Check your sketch by plotting the function on DERIVE.

(v)   Repeat (iv) with other functions of the form $m\sin(nx)$ for different values of m and n.

(C)   (i)   **Author** and **Simplify** VECTOR(sin(x+n),n,0,1.5,0.5).

          Plot with the scales $x:\dfrac{\pi}{2}$, $y:1$.

          In which direction are successive graphs being translated?
          Move the cross onto the x axis. **Zoom In** on the values to the left of the
          origin where successive graphs cross the x axis. How do these values
          relate to the numbers you see in the vector?
          Delete the graphs and remove the expressions then repeat the procedure
          for VECTOR(sin(x–n),n,0,1.5,0.5).
          Zoom In on the values to the right of the origin where successive graphs
          cross the x axis.

     (ii) Repeat with VECTOR(cos(x–n),0,1.5,0.5).
          Zoom In on the peaks to the left and right of the origin.

     (iii) Repeat with tan(x+n) and tan(x–n).

     (iv) Repeat with sin(2x+n) for **Scale** $x:\dfrac{\pi}{4}$, $y:1$ and sin(4x–n) for **Scale** $x:\dfrac{\pi}{8}$,
          $y:1$.

(D)   (i)   **Author** and **Simplify** VECTOR$(\sin x+n,n,0,3)$.

          Plot with **Scale** $x:\dfrac{\pi}{2}$, $y:4$. In which direction are the graphs translated?

          Move the cross to the y axis. **Zoom In** on those values where successive
          graphs cross the y axis. How do these values relate to the numbers in the
          vector?

          Delete the graphs and remove the expressions then repeat with
          VECTOR$(\sin x-n,n,0,3)$.

     (ii) Repeat with cosx + n.

     (iii) Repeat with tanx – n.

     These four activities have shown that the graphs of the trigonometric functions
sine, cosine and tangent may be subtly altered by including numbers in various places
within the function description. Particularly important are the sine and cosine
functions which are often used to model physical situations involving oscillations.

Consider the sine function of time t

$$f(t) = a\sin(wt + \alpha) + c$$

which may be regarded as typical of them all.

The four parameters, a, w, α and c each cause a specific alteration in the graph of $\sin t$. Taking them in the order of the investigations

1. a is the *amplitude* of the function and affects the size of the oscillations.

2. w is the *angular frequency* of the function and affects the number of oscillations.

3. α is related to a *horizontal translation* which is determined by solving for t the

equation $wt + \alpha = 0$ ie. $t = -\dfrac{\alpha}{w} = t_0$; the graph is translated to the left if $\alpha > 0$

and to the right if $\alpha < 0$.

4. c is a *vertical translation* which is upwards if $c > 0$ and downwards if $c < 0$.

There are two further related concepts to consider. We saw in section 3.5 that the period of the fundamental sine function is 2π. Since the function $\sin 2t$ has twice as many oscillations, its period must be $2\pi/2 = \pi$. Likewise the function $\sin 3t$ has three times as many oscillations and so its period must be $2\pi/3$.

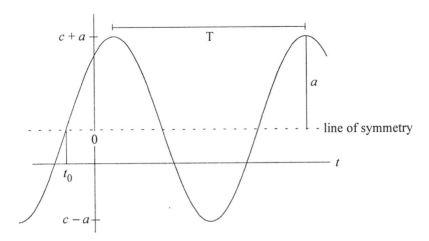

Figure 3.15   Properties of $a\sin(wt+\alpha) + c$

In general the *period*, T, of a trigonometric function with angular frequency w

is given by $T = \dfrac{2\pi}{w}$. Thus we see that the higher the angular frequency the shorter is

the period.

The word "period" suggests an interval of time (which is why we have written f in terms of t) and is appropriate to functions which repeat themselves in cycles. It can be important to know how much of the cycle takes place in a given unit of time. This is known as *frequency*, and in practical circumstances is typically measured in cycles per second or Hertz (Hz).

In general the frequency, f, of a trigonometric function of period T will be given by

$f = \dfrac{1}{T}$. Thus

$$f = \frac{w}{2\pi}$$

or $\qquad w = 2\pi f$

which shows that the angular frequency is directly proportional to the frequency.

Exercise 3D

1.   State (i) the amplitude, (ii) the angular frequency, (iii) the horizontal translation and (iv) the vertical translations of the following expressions.

(a) $3\sin 5t$ (b) $4\cos\left(t - \dfrac{\pi}{2}\right)$ (c) $\cos(3t+2\pi)$

(d) $\sin(3t) + 2$ (e) $\sin(3t+2)$ (f) $2\sin(4t-\pi) + 1$

(g) $\cos(\tfrac{1}{2}t-\pi) - 3$ (h) $0.1\sin(2\pi t-3\pi) + 0.5$

2.   Sketch (by hand) the graphs of the expressions in problem 1. (Check your answers using DERIVE).

3.   Write down:

(a)   the cosine function with amplitude 6 and angular frequency 2

(b)   the sine function with amplitude 3, angular frequency 4 and horizontal

translation $-\dfrac{\pi}{4}$

(c)   a tangent function with angular frequency 5 and vertical translation 3

(d)   the sine function with amplitude 10, angular frequency 5 and horizontal

translation $\dfrac{\pi}{10}$

(e)   the cosine function with amplitude 0.5, angular frequency 3, horizontal

translation $\dfrac{2}{3}$ and vertical translation -1

(f)   the period of a trigonometric function with angular frequency 20

(g)   the frequency of a trigonometric function with period 0.1π

(h)   the angular frequency of a trigonometric function with frequency
        100 Hz.

4.   Use DERIVE to plot the functions (a) to (e) in problem 3.

5.   State (i) the period, (ii) the frequency of the following functions.

(a) $\cos(3t-1)$ (b) $\sin(5t+\pi)$ (c) $4\sin(8\pi t+1)$

(d) $10\cos(0.5\pi t-3\pi)$ (e) $\sin\left(8t-\dfrac{\pi}{2}\right)$

6.   Sketch (by hand) the graphs of the functions in problem 5. (Check your
        answers using DERIVE.

3.8   MODELLING WITH TRIGONOMETRIC FUNCTIONS

In this section we shall apply the concepts of amplitude, frequency, period, etc to physical phenomena.

Example 3C

The following table represents the number of hours of daylight in the city of Plymouth over two years. The measurements began on 15th April 1990. Express the number of hours of daylight N as a function of the number of days t since 15th April 1990.

t	0	91.25	182.5	273.75	365	456.25	547.5	638.75	730
N	14	19	14	9	14	19	14	9	14

Solution

We are looking for a function of the form $N = a\sin(wt + \alpha) + c$. We see that the largest and smallest values of N are $N = 9$ and $N = 19$, so the amplitude of the oscillation $a = \frac{1}{2}(19 - 9) = 5$ and the vertical translation $c = 14$.

The period of the phenomenon is 365 days which gives an angular frequency of

$w = \dfrac{2\pi}{365} = 0.0172$. The horizontal shift is zero because when $t = 0$, $N = 14$ the mean value of the oscillation which gives

$$5\sin\alpha + 14 = 14$$
$$5\sin\alpha = 0$$

hence

$$\alpha = 0 .$$

Thus the model for the number of daylight hours as a function of time t is

$$N = 5\sin(0.0172t) + 14 .$$

Example 3D

The blood pressure, P millibars (mb), of a patient in hospital is modelled by the function

$$P = 25\cos(6t) + 95$$

where t is time measured in seconds.

Use this model to find the maximum (systolic) and minimum (diastolic) pressures. What is the period of time between consecutive systolic pressures? What is the patient's blood pressure after 2 seconds?

Solution

The amplitude of the function, $a = 25$, will be attained whenever $\cos(6t) = 1$ or $\cos(6t) = -1$.

Thus the systolic pressure is $25 + 95 = 120$ mb and the diastolic pressure is $-25 + 95 = 70$mb.

Consecutive maxima (or minima) are repeated in a time equal to the period of the oscillation ie. $\dfrac{2\pi}{6} = 1.047\,\text{s}$.

When $t = 2s$ (remembering to work in radians) the pressure is given by

$$P = 25\cos(12) + 95 = 116.096 \text{ mb (to 3 dp).}$$

These examples show how we can use the functions sine and cosine to model any system which behaves in a wave like motion. All that is required is either a graph showing the motion of the system or data concerning the period, amplitude and frequency. The functions sine and cosine are often called *sinusoidal functions*.

Exercise 3E

1.   In the USA, electricity is supplied at a frequency of 60 Hz. What function
        would model the current across a resistor if the current's amplitude was 80 mA?

2.   A mass attached to the end of a spring oscillates so that its displacement s cm
        from a central position as given by

$$s = 6\sin(3\pi t)$$

where t is the time in seconds. What is the maximum displacement from the
central position? What is the frequency of the oscillation? Where is the object

(a)   5 seconds into the motion?

(b)   17.3 seconds into the motion?

3.   The mean monthly temperature in Benidorm peaks in August at 27°C and has
        a minimum in February of 5°C. Assuming that the variation in temperature is
        sinusoidal, obtain a mathematical model to represent the mean monthly
        temperature. Use this model to predict the mean monthly temperature in June
        and in January.

4.   The Bristol Channel is known to have an unusually high rise and fall in tide
        levels. For a typical spring tide there is as much as a 13m difference between
        high tide and low tide. Given that there is a gap of 12.4 hours between
        successive high tides, find the values of a and w in the model

$$h = a\sin(wt+\alpha)$$

which gives the height h in metres of the water above mean sea level as a
function of time t in hours.

On a given day, at the time of a spring tide, high water occurs at midnight
($t = 0$). Find the value of α and use the model to determine the height above
mean sea level of the water at midnight on the following three successive
nights.

3.9   SOLVING TRIGONOMETRIC EQUATIONS

In this section we shall see that solving trigonometric equations is a quite different process to solving polynomial equations.

DERIVE ACTIVITY 3D

(A)   (i)   **Author** and **Plot** sin x using **Scale** x:π y:1.

        (ii)   **Author** and **Plot** 0.4 on the same grid.
                How many intersection points of the two graphs can you see?
                How many more intersection points do you think there are?

        (iii)   Use the arrow keys to find some of the solutions of the equation
                sin $x = 0.4$.

(B)   Repeat (A) using cos x.

(C)   Repeat (A) using tan x.

From this activity we can see that even the simplest kinds of trigonometric equation,

$$\sin x = 0.4, \quad \cos x = 0.4, \quad \tan x = 0.4$$

possesses infinitely many solutions. This is because the trigonometric functions sin, cos and tan are periodic.

Exercise 3F

Use your calculator in the degree mode to solve

1.   sin $x = 0.4$ 2.   cos $x = 0.4$ 3.   tan $x = 0.4$

4.   sin $x = -0.8$ 5.   cos $x = -0.8$ 6.   tan $x = -0.8$.

The results obtained in Exercise 3F show that an electronic calculator will supply only *one* of the infinitely many possible answers to each problem. The reason for this will be explained more fully in section 3.10.

Figure 3.16 shows the graphs of $y = \sin x$ and $y = 0.4$ for $-540° \leq x \leq 540°$.

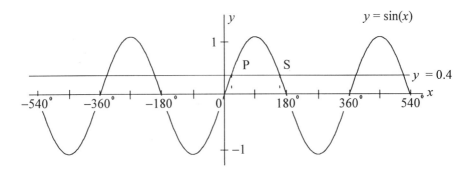

Figure 3.16

When solving the equation $\sin x = 0.4$ with a calculator the answer $x = 23.6°$ is obtained. In figure 3.16 this solution is labelled p because we shall call it the *primary solution* of the equation.

Because of the symmetry of the sine graph the next solution nearest to p is $180 - 23.6° = 156.4°$. In figure 3.16 this is labelled s because we shall call it the *secondary solution* of the equation.

These solutions, p and s, together with the periodic nature of the sine graph will supply all possible solutions to the equation $\sin x = 0.4$. Since we know that the period of the sine graph is $360°$ we can now see that the general solution of the equation $\sin x = 0.4$ must be

$$x = 360°n + 23.6°$$

or

$$x = 360°n + 156.4°$$

for $n = 0, \pm1, \pm2, \pm3, \ldots$

Example 3E

Solve $\sin x = -0.73$ for $-720° \leq x \leq 720°$.

Solution

Using a calculator we find the primary solution $p = -46.9°$. The secondary solution is given by $s = -180° + 46.9° = -133.1°$. So the general solution is given by

$$x = 360°n - 46.9° \quad \text{or} \quad x = 360°n - 133.1° \,.$$

Taking $n = 0$ we have the primary and secondary solutions

$$x = -46.9° \quad \text{and} \quad x = -133.1° \,.$$

Taking $n = 1$ we obtain

$$x = 313.1° \quad \text{and} \quad x = 226.9° \,.$$

Taking $n = -1$ we obtain

$$x = -406.9° \quad \text{and} \quad x = -493.1° \,.$$

Taking $n = 2$ we obtain

$$x = 673.1° \quad \text{and} \quad x = 586.9° \,.$$

Taking any more values of n would take us outside the required interval $-720° \leq x \leq 720°$. Thus the solutions in this interval are

$$x = -493.1°, -406.9°, -133.1°, -46.9°, 226.9°, 313.1°, 586.9°, 673.1° \,.$$

Figure 3.17 displays these solutions graphically.

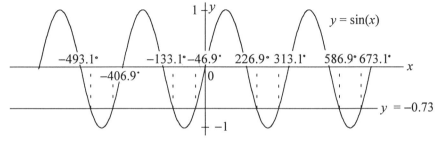

Figure 3.17

A similar procedure may be adopted for equations involving cos and tan.

Example 3F

Solve $\cos x = 0.4$ for $-360° \le x \le 360°$.

Solution

The primary solution given by a calculator is $p = 66.4°$.

In the case of the cosine graph the secondary solution is $s = -66.4°$ as shown in figure 3.18.

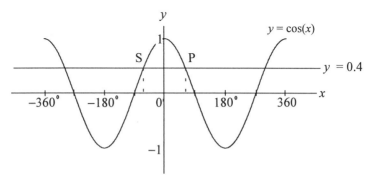

Figure 3.18

The period of the cosine graph is 360° and therefore we see with reference to figure 3.18 that the general solution of $\cos x = 0.4$ must be

$$x = 360°n \pm 66.4° .$$

When $n = 0$ we get $x = 66.4°$ and $-66.4°$.
When $n = 1$ we get $x = 293.6°$ and $-293.6°$.

Any further values of n will take us outside the range $-360° \le x \le 360°$.

So the solutions in this interval are

$$x = -293.6°, -66.4°, 66.4°, 293.6°.$$

Example 3G

Solve $\tan x = -0.8$ for $-450° \le x \le 450°$.

Solution

The primary solution $p = -38.7°$. Figure 3.19 shows that there is no secondary solution for simple tangent equations.

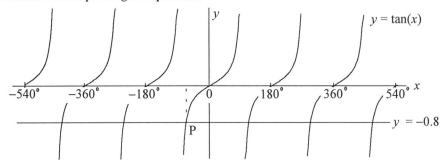

Figure 3.19

The period of the tangent function is $180°$ and we can see from figure 3.19 that the general solution of $\tan x = -0.8$ is

$$x = 180°n + (-38.7°) .$$

Thus

when $n = $ 0, $x = -38.7°$
when $n = $ 1, $x = 141.3°$
when $n = -1$, $x = -218.7°$
when $n = $ 2, $x = 321.3°$
when $n = -2$, $x = -398.7°$
when $n = $ 3, $x = 501.3°$

Any further value of n takes us outside the interval $-540° \le x \le 540°$. So we have the required solutions

$$x = -398.7°, -218.7°, -38.7°, 141.3°, 321.3°, 501.3°.$$

Exercise 3G

1.   For $-360° \leq x \leq 360°$ solve the equations

   (a) $\sin x = 0.47$ (b) $\cos x = -0.17$ (c) $\tan x = 1.4$
   (d) $\cos x = 0.68$ (e) $\tan x = -2$ (f) $\sin x = -0.89$

2.   Use DERIVE to display your answers to problem 1 graphically.

3.   For $-180° \leq \theta \leq 540°$ solve the equations

   (a) $\cos \theta = 0.31$ (b) $\tan \theta = -0.89$ (c) $\sin \theta = 0.9$.

4.   Use DERIVE to display your answers to problem 2 graphically.

3.10   INVERSE TRIGONOMETRIC FUNCTIONS

Section 3.9 showed us that trigonometric equations have infinitely many solutions and you saw how to state general solutions for simple sine equations, cosine equations and tangent equations.

We noted that our electronic calculators supply only one of the infinitely many possible solutions, and we called this the primary solution. Graphical considerations enabled us to find a secondary solution and then an expression for a general solution.

DERIVE ACTIVITY 3E

(A)   (i)   **Author** and **Plot** ASIN(x) using the **Scale** x:1 y:π/2.
      (ii)   **Manage Substitute** $x = 0.4$ and **approX** to obtain an answer in radians.
      (iii)   Convert this answer to degrees.

(B)   Repeat (A) using ACOS(x) and ATAN(x).

(C)   Compare parts (ii) and (iii) of each of these activities with your calculator solutions to the equations

      (i) $\sin x = 0.4$ (ii) $\cos x = 0.4$ (iii) $\tan x = 0.4$

using first radian mode and then degree mode.

It would be impractical for your calculator or computer to deliver infinitely many answers to an equation. When we find the *inverse* of a trigonometric function it is necessary to restrict the range of possible values so that only one value, usually the smallest or most convenient, is selected. We can then determine what the other solutions will be through the methods demonstrated in section 3.9. These restricted ranges of possible values are called *principal values* and they provide the primary solution to any trigonometric equation.

Notations for the inverse trigonometric functions vary. The most usual are

$$y = \sin^{-1} x \qquad y = \arcsin x \qquad y = \text{ASIN}\, x$$
$$y = \cos^{-1} x \qquad y = \arccos x \qquad y = \text{ACOS}\, x$$
$$y = \tan^{-1} x \qquad y = \arctan x \qquad y = \text{ATAN}\, x$$

Figure 3.20, figure 3.21 and figure 3.22 show the graphs of the inverse trigonometric functions and their principal values.

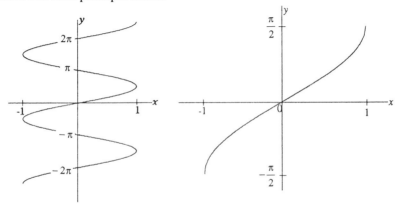

Figure 3.20   Graph of $y = \arcsin x$

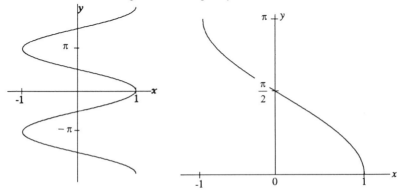

Figure 3.21   Graph of $y = \arccos x$

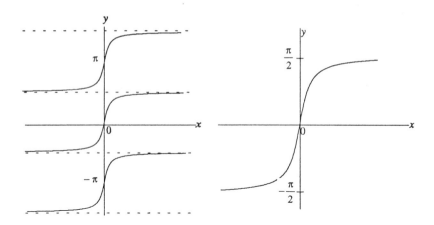

Figure 3.22   Graph of arctan x

Exercise 3H

1.   Using your calculator find the principal value in degrees of each of the
     following.

     (a) $\cos^{-1}(0.62)$ (b) $\tan^{-1}(-1.05)$ (c) $\sin^{-1}(-0.88)$

     (d) $\tan^{-1}(6.59)$ (e) $\sin^{-1}(0.06)$ (f) $\cos^{-1}(-0.95)$

2.   Use DERIVE to find the principal value in radians of each of the following.

     (a)   ATAN(–3.5)   (b)   ASIN(–0.1)   (c)   ACOS(0.06)

     (d)   ACOS(–0.33)   (e)   ATAN(0.5)   (f)   ASIN(0.11)

3.   Find the values of each of the following expressions.

     (a) $\sin(\cos^{-1}0.1)$ (b) $\cos(\sin^{-1}0.1)$ (c) $\tan(\cos^{-1}0.8)$

     (d) $\tan(\sin^{-1}-0.3)$ (e) $\cos(\tan^{-1}3.2)$ (f) $\sin(\tan^{-1}-1.6)$

     (g) $\sin(\sin^{-1}0.8)$ (h) $\cos(\cos^{-1}-0.55)$ (i) $\tan(\tan^{-1}0.75)$

DERIVE ACTIVITY 3F

(A)   (i)   **Author** and **Plot** sin x using **Scale** x:π y:1.

       (ii)   **Author** and **soLve** sin $x = 0.4$. Three values will be displayed. Use
              approX on each of these to obtain $x = -3.55310$, $x = 2.73007$ and
              $x = 0.411516$.

       (iii) **Author** and **Plot** in succession [$-3.5531,y$], [$2.73007,y$] and
              [$0.411516,y$].

              Three vertical lines are superimposed on the graph of $y = \sin x$. The last
              one plotted locates the primary solution of the equation sin $x = 0.4$ while
              the second one plotted locates the secondary solution.

(B)   Repeat (A) to find the primary and secondary solutions of cos $x = 0.65$.

(C)   Repeat (A) to find the primary solution of tan $x = -1.5$.

In (B) the primary solution is the first to be generated and the secondary one is
generated next, $x = 0.863211$ followed by $x = -0.863211$.

In (C) only the primary solution is produced $x = -0.982793$.

This activity shows that you can use the **soLve** command in DERIVE when finding
the solution of trigonometric equations. However, care must be exercised to choose
the correct primary solution because initially a number of expressions may be
produced in exact form using the inverse functions ASIN, ACOS and ATAN.

Exercise 3I

Use the soLve command in DERIVE and find all solutions of the following equations
for $-2\pi \leq t \leq 2\pi$ Give your answers in radians.

1.   cos $t = 0.47$ 2.   sin $t = 0.163$ 3.   tan $t = -0.09$

4.   sin $t = -0.14$ 5.   tan $t = 1.3$ 6.   cos $t = 0.01$

3.11 TRIGONOMETRIC IDENTITIES

You will have already seen that there are some simple relationships between sine, cosine and tangent. Often when solving a problem it can be necessary to change the form of a trigonometric expression in order to progress. In this section we will consider a number of other important relationships often known as *identities*.

The first result, sometimes referred to as the *Pythagorean Relationship*, can be deduced by considering the right angle triangle shown in figure 3.23.

We know that

$$a^2 + b^2 = c^2$$

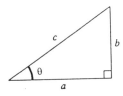

which can be rewritten as

$$\frac{a^2}{c^2} + \frac{b^2}{c^2} = 1$$

Figure 3.23

or

$$\left(\frac{a}{c}\right)^2 + \left(\frac{b}{c}\right)^2 = 1 .$$

But from figure 3.23 $\sin\theta = \dfrac{b}{c}$ and $\cos\theta = \dfrac{a}{c}$, so the equation above becomes

$$(\sin\theta)^2 + (\cos\theta)^2 = 1$$

which is usually written as

$$\sin^2\theta + \cos^2\theta = 1 .$$

This is a very important result, known as an *identity*, and one that you will encounter frequently in your studies.

It is also useful to introduce three new functions which are defined as the reciprocals of the basic three trigonometric functions. They are

$$secant \qquad \sec\theta = \frac{1}{\cos\theta} \qquad\qquad cosecant \qquad \cosec\theta = \frac{1}{\sin\theta}$$

$$cotangent \qquad \cot\theta = \frac{1}{\tan\theta} = \frac{\cos\theta}{\sin\theta} .$$

With these new functions and the Pythagorean relationship we can solve more complicated looking trigonometric equations.

Example 3H

Find the solution of the equation

$$\cos\theta - \sec\theta = \tan\theta ,$$

for $0 \le \theta < 360°$.

Solution

The first step is to express sec and tan in terms of sin and cos to give

$$\cos\theta - \frac{1}{\cos\theta} = \frac{\sin\theta}{\cos\theta} .$$

Now multiplying by $\cos\theta$ gives

$$\cos^2\theta - 1 = \sin\theta .$$

Using $\sin^2 + \cos^2\theta = 1$ (in the form $\cos^2\theta = 1 - \sin^2\theta$) leads to

$$1 - \sin^2\theta - 1 = \sin\theta$$

or

$$\sin\theta + \sin^2\theta = 0 .$$

Factorising gives

$$\sin\theta(1+\sin\theta) = 0$$

so

$$\sin\theta = 0 \quad \text{or} \quad \sin\theta = -1$$

so

$$\theta = 0°, \ 180° \text{ or } 270°.$$

It is possible to form other identities in terms of sec, cosec and cot from the identity obtained above. For example starting with

$$\sin^2\theta + \cos^2\theta = 1$$

we can divide all terms by $\cos^2\theta$ to obtain

$$\frac{\sin^2\theta}{\cos^2\theta} + \frac{\cos^2\theta}{\cos^2\theta} = \frac{1}{\cos^2\theta}$$

or

$$\tan^2\theta + 1 = \sec^2\theta .$$

Example 3I

Prove that

$$\tan^2\theta - \sin^2\theta = \tan^2\theta \sin^2\theta .$$

Solution

Starting with the LHS and expression $\tan\theta$ in terms of $\sin\theta$ and $\cos\theta$ gives

$$\tan^2\theta - \sin^2\theta = \frac{\sin^2\theta}{\cos^2\theta} - \sin^2\theta$$

$$= \sin^2\theta\left(\frac{1}{\cos^2\theta} - 1\right)$$

$$= \sin^2\theta(\sec^2\theta - 1) .$$

But we know the identity $\tan^2\theta + 1 = \sec^2\theta$, so $\tan^2\theta = \sec^2\theta - 1$. Using this gives

$$\sin^2\theta(\sec^2\theta - 1) = \sin^2\theta \tan^2\theta ,$$

to complete the proof.

Exercise 3J

1.   In DERIVE plot the graphs of $\cos^2 x$, $\sin^2 x$ and $\sin^2 x + \cos^2 x$. Describe the relationships that exist between the three curves.

2.   (a)   In DERIVE plot the graphs of $\cos x$ and $\sec x$.

     (b)   Sketch graphs of $\sin x$ and $\operatorname{cosec} x$. Then check your result with DERIVE.

     (c)   Repeat (ii) for $\tan x$ and $\cot x$.

3.   Solve the equations below, giving all solutions in the range $0° \le x \le 360°$.

     (a) $\sin^2 x + 3\cos^2 x = 2$ (e) $4\sin x - 2\operatorname{cosec} x = \cot x$

     (b) $3\cos x - 2\sec x = 3\tan x$ (f) $5\sin^2 x + \sin x - 4 = 3\cos^2 x$

     (c) $\sin x + 4 = 7\cos^2 x$ (g) $2 + \cos^2 x = 3\sin x \cos x$

     (d) $6\sin^2 x - 4 = \cos x$ (h) $2\cot x = 3\tan x$

4.   Prove that

$$1 + \cot^2 \theta = \operatorname{cosec}^2 \theta .$$

5.   Prove that

$$\cos^2 \theta - \sin^2 \theta = 2\cos^2 \theta - 1 ,$$

     hence show that

$$\cos^4 \theta - \sin^4 \theta = 2\cos^2 \theta - 1 .$$

6.   Prove each of the identities given below.

     (a) $\sin^2 \theta - \cos^2 \theta = 2\sin^2 \theta - 1$ (d) $\cos\theta(\cot\theta + \tan\theta) = \operatorname{cosec}\theta$

     (b) $(1 - \cos\theta)(1 + \cos\theta) = \sin^2 \theta$ (e) $\sec^2 \theta(\sec^2 \theta - 1) = \tan^2 \theta\,(1 + \tan^2 \theta)$

     (c) $(\sin\theta + \cos\theta)^2 + (\sin\theta - \cos\theta)^2 = 2$ (f) $\sin^2 A - \sin^2 B = \cos^2 B - \cos^2 A$

3.12 FURTHER IDENTITIES

There are a number of further identities that will not be proved in this book, but which are of importance in mathematics. The first group of these that we will consider are *Sum* and *Difference Formulae*, which are listed below.

$$\sin(A+B) = \sin A \, \cos B + \cos A \, \sin B$$

$$\sin(A-B) = \sin A \, \cos B - \cos A \, \sin B$$

$$\cos(A+B) = \cos A \, \cos B - \sin A \, \sin B$$

$$\cos(A-B) = \cos A \, \cos B + \sin A \, \sin B$$

$$\tan(A+B) = \frac{\tan A + \tan B}{1 - \tan A \tan B}$$

$$\tan(A-B) = \frac{\tan A - \tan B}{1 + \tan A \tan B}$$

It is a very simple matter to obtain a further set of identities known as the *Double Angle Formulae* from the previous set of identities, which give results for $\sin(2A)$, $\cos(2A)$ and $\tan(2A)$.

For $\sin(2A)$ the result is obtained as below:

$$\begin{aligned}
\sin(2A) &= \sin(A+A) \\
&= \sin A \cos A + \cos A \sin A \\
&= 2 \sin A \cos A .
\end{aligned}$$

For $\cos(2A)$:

$$\begin{aligned}
\cos(2A) &= \cos(A+A) \\
&= \cos A \cos A - \sin A \sin A \\
&= \cos^2 A - \sin^2 A .
\end{aligned}$$

By using the identity $\cos^2 A + \sin^2 A = 1$, this identity can also be written as

$$\cos(2A) = 2\cos^2 A - 1$$

or

$$\cos(2A) = 1 - 2\sin^2 A .$$

DERIVE ACTIVITY 3G

(A) (i)   **Author** and **Plot** $2\sin x \cos x$.
           What are the amplitude and period of this wave?
           What sine wave would have this amplitude and period?
           Author and **Plot** your answer to check.

     (ii) **Transfer Clear** the screen and then **Author** and **Plot**, $\cos^2 x - \sin^2 x$,
           $2\cos^2 x - 1$ and $1 - 2\sin^2 x$. What are the amplitude and period of these
           waves? What cosine wave would have this amplitude and period?
           Author and **Plot** your answer to check.

     (iii) **Author** and **Plot** $3\sin x - 4\sin^3 x$. What single sine wave is equivalent to
           the wave you have on the screen? Check your answer by Authoring and
           Plotting it.

(B) (i)   **Author** and **Plot** $\sin(5x) + \sin(3x)$.
           Also **Author** and **Plot** $2\sin(4x)\cos(x)$. Comment on your results.

     (ii) Repeat (i) for $\sin(9x) + \sin(3x)$ and $2\sin(6x)\cos(3x)$.

     (iii) Express $\sin(7x) + \sin(3x)$ in the form $2\sin(Ax)\cos(Bx)$.
           Check your result by plotting both expressions.

     (iv) By plotting compare $\sin(10x) - \sin(4x)$ and $2\cos(7x)\sin(3x)$.
           Try to predict an alternative form for $\sin(12x) - \sin(8x)$. Check your
           result by plotting on DERIVE.

The next set of identities are known as the *Factor Formulae* and are listed below.

$$\sin A + \sin B = 2\sin\left(\frac{A+B}{2}\right)\cos\left(\frac{A-B}{2}\right)$$

$$\sin A - \sin B = 2\cos\left(\frac{A+B}{2}\right)\sin\left(\frac{A-B}{2}\right)$$

$$\cos A + \cos B = 2\cos\left(\frac{A+B}{2}\right)\cos\left(\frac{A-B}{2}\right)$$

$$\cos A - \cos B = -2\sin\left(\frac{A+B}{2}\right)\sin\left(\frac{A-B}{2}\right).$$

Example 3J

For the triangles in figure 3.24, show that

(i) $\cos(A+B) = \dfrac{33}{65}$

(ii) $\sin(A+B) = \dfrac{56}{65}$

(iii) $\tan(A+B) = \dfrac{56}{33}$.

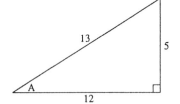 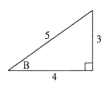

Figure 3.24

Solution

Using the sides of the first triangle

$$\cos A = \frac{12}{13}, \quad \sin A = \frac{5}{13}, \quad \tan A = \frac{5}{12},$$

and from the second triangle

$$\cos B = \frac{4}{5}, \quad \sin B = \frac{3}{5}, \quad \tan B = \frac{3}{4}.$$

(i)   Using the addition formula for $\cos(A+B)$

$$\cos(A+B) = \cos A \cos B - \sin A \sin B$$

$$= \frac{12}{13} \times \frac{4}{5} - \frac{5}{13} \times \frac{3}{5} = \frac{33}{65}.$$

(ii)   From the addition formula for $\sin(A+B)$

$$\sin(A+B) = \sin A \cos B + \cos A \sin B$$

$$= \frac{5}{13} \times \frac{4}{5} + \frac{12}{13} \times \frac{3}{5} = \frac{56}{63}.$$

(iii)   From the addition formula for $\tan(A+B)$

$$\tan(A+B) = \frac{\tan A + \tan B}{1 - \tan A \tan B}$$

$$= \frac{\dfrac{5}{12} + \dfrac{3}{4}}{1 - \dfrac{5}{12} \times \dfrac{3}{4}} = \frac{\dfrac{7}{6}}{\dfrac{33}{48}}$$

$$= \frac{56}{33} \ .$$

Example 3K

Prove that
$$\sin(A+B)\sin(A-B) = \sin^2 A - \sin^2 B \ .$$

Solution

Using the sum and difference formulae

$$\sin(A+B)\sin(A-B) = (\sin A \cos B + \cos A \sin B)(\sin A \cos B - \cos A \sin B)$$

$$= \sin^2 A \cos^2 B - \sin A \sin B \cos A \cos B$$
$$\qquad + \sin A \sin B \cos A \cos B - \cos^2 A \sin^2 B$$
$$= \sin^2 A \cos^2 B - \cos^2 A \sin^2 B$$
$$= \sin^2 A (1 - \sin^2 B) - (1 - \sin^2 A) \sin^2 B$$
$$= \sin^2 A - \sin^2 A \sin^2 B - \sin^2 B + \sin^2 A \sin^2 B$$
$$= \sin^2 A - \sin^2 B$$

Example 3L

Prove that $\tan(x+45°) = \dfrac{1+\tan x}{1-\tan x}$.

Solution

Using the appropriate addition formula gives

$$\tan(x+45°) = \frac{\tan x + \tan 45°}{1 - \tan x \tan 45°}$$
$$= \frac{\tan x + 1}{1 - \tan x} \qquad \text{(since } \tan 45° = 1\text{)}.$$

Example 3M

Prove that

$$\frac{\sin 10A - \sin 6A}{\cos 10A + \cos 6A} = \tan 2A$$

Solution

Applying the appropriate factor formula to the top and bottom of the LHS gives

$$\frac{\sin 10A - \sin 6A}{\cos 10A + \cos 6A} = \frac{2\cos\left(\dfrac{10A+6A}{2}\right)\sin\left(\dfrac{10A-6A}{2}\right)}{2\cos\left(\dfrac{10A+6A}{2}\right)\cos\left(\dfrac{10A-6A}{2}\right)}$$
$$= \frac{2\cos(8A)\sin(2A)}{2\cos(8A)\cos(2A)}$$
$$= \frac{\sin(2A)}{\cos(2A)}$$
$$= \tan(2A) \ .$$

Exercise 3K

1.   For the triangle in figure 3.25, find $\cos(A+B)$, $\sin(A+B)$ and $\tan(A+B)$.

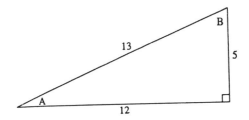

Figure 3.25

Comment on your results.

2.   If $\cos A = \dfrac{24}{25}$ and $\cos B = \dfrac{8}{17}$ find

(a) $\sin A$ (e) $\sin(A-B)$ (i) $\cos(2A)$

(b) $\sin B$ (f) $\tan(B-A)$ (j) $\sin(2B)$

(c) $\tan A$ (g) $\cos(A+B)$

(d) $\tan B$ (h) $\tan(A+B)$.

3.   Repeat problem 2 given that $\cos A = 0.5$ and $\cos B = 0.8$.

4.   (a)   Show that $\sin(A + 90°) = \cos A$.

      (b)   Find simpler expressions for

            (i) $\sin(A + 180°)$ (iv) $\cos(A + 180°)$

            (ii) $\sin(A - 90°)$ (v) $\cos(A - 90°)$

            (iii) $\cos(A + 90°)$ (vi) $\tan(A - 45°)$.

5.   Find alternative expressions for

      (a) $\cos 8x \sin x - \sin 8x \cos x$ (e) $\sin 40° \sin 10°$

      (b) $\sin 60° + \sin 20°$ (f) $\sin 69° \cos 40° - \sin 40° \cos 69°$

      (c) $\cos 50° - \cos 30°$ (g) $\sin 45° - \sin 35°$

      (d) $\sin 50° \cos 30°$ (h) $\sin 45° \cos x + \cos 45° \sin x.$

6.   (a)   Prove that $\sin(3A) = 3\sin A - 4\sin^3 A$.

      (b)   Find a similar result for $\cos(3A)$.

      (c)   Express $\sin(4A)$ in terms of $\sin A$ and $\cos A$.

7.   Prove that

      (a) $\cos(A+B)\cos(A-B) = \cos^2 A - \sin^2 B$

      (b) $\cos(A+B) + \cos(A-B) = 2\cos A \cos B$

      (c) $\dfrac{\sin A}{\sin B} + \dfrac{\cos A}{\cos B} = \dfrac{2\sin(A+B)}{\sin(2B)}$

      (d) $\cos(3\theta) - \cos(7\theta) = 2\sin(5\theta)\sin(2\theta)$.

8.   Show that when two wave forms of roughly the same amplitude but slightly
      different periods are added together the result is a wave of roughly the same
      period but with an amplitude that varies with a frequency equal to the
      difference in frequencies of the two waves.

3.13 THE EXPRESSION $A\sin x + B\cos x$

DERIVE ACTIVITY 3H

(A) (i)   In DERIVE **Author** and **Plot** $\sin(x) + \cos(x)$. Use **Scale** $x{:}\pi$ and $y{:}1$.

(ii)   What is the amplitude of the wave that you produce? Call this value R.

(iii) How far to the left of the y-axis does the curve cross the x-axis? Call this value α.

(iv)   Now **Author** and **Plot** $R\sin(x+\alpha)$ using your values of R and α. What do you notice?

(B)   Repeat (A) for each combination given below.

(i) $2\sin(x) + 2\cos(x)$

(ii) $3\sin(x) + 4\cos(x)$

(iii) $5\sin(x) + 12\cos(x)$

(iv) $\sin(x) - \cos(x)$

(C)   Repeat (A) for each combination given below.

(i) $3\sin(4x) + 4\cos(4x)$

(ii) $5\cos(3x) + 12\sin(3x)$

(iii) $\sin(2x) - 2\cos(2x)$

(iv) $2\cos(5x) - 5\sin(5x)$

What do you deduce about the amplitude of the wave $A\sin(wx) + B\cos(wx)$?

Simplifying $A\sin x + B\cos x$

From the DERIVE Activity you will have seen that $A\sin x + B\cos x$ produces a sine wave of the form $R\sin(x+\alpha)$. Using the addition formulae it is possible to find relationships between the constants R and α and the constants A and B.

Using the addition formula

$$R\sin(x+\alpha) = R\sin x\cos\alpha + R\cos x\sin\alpha$$
$$= R\cos\alpha\sin x + R\sin\alpha\cos x .$$

But we know that

$$R\sin(x+\alpha) = A\sin x + B\cos x .$$

Comparing the right hand sides of the above equations gives

$$R\cos\alpha = A \quad \text{and} \quad R\sin\alpha = B .$$

To obtain an expression for R, begin by squaring and adding A and B, to give

$$A^2 + B^2 = R^2\cos^2\alpha + R^2\sin^2\alpha$$
$$= R^2(\cos^2\alpha + \sin^2\alpha)$$
$$= R^2$$

so

$$R = \sqrt{A^2 + B^2}$$

To obtain an expression for α, divide B by A to give

$$\frac{B}{A} = \frac{R\sin\alpha}{R\cos\alpha} = \frac{\sin\alpha}{\cos\alpha} = \tan\alpha$$

so

$$\alpha = \tan^{-1}\left(\frac{B}{A}\right).$$

Example 3N

(i)   Write $4\sin x + 3\cos x$ in the form $R\sin(x+\alpha)$.

(ii)   Find the solutions of the equation

$$4\sin x + 3\cos x = 2.5 \, ,$$

that lie in the range $0 \le x \le 360°$.

Solution

(i)   For $4\sin x + 3\cos x$ we have $A = 4$ and $B = 3$, so

$$R = \sqrt{4^2 + 3^2} = \sqrt{25} = 5 \quad \text{and} \quad \alpha = \tan^{-1}\left(\frac{3}{4}\right) = 36.9° \text{ or } 216.9°.$$

To decide which value of α to use, note that $\cos\alpha = \dfrac{4}{5}$ and $\sin\alpha = \dfrac{3}{5}$. In this case both are positive so $0° < \alpha < 90°$. Hence α must take the value of $36.9°$.

(ii)   The equation

$$4\sin x + 3\cos x = 2.5$$

can be rewritten as

$$5\sin(x + 36.9°) = 2.5$$
or
$$\sin(x + 36.9°) = 0.5 \, .$$

The solutions are then given by

$$x + 36.9° = 30°, \ 150°, \ 390°, \text{ etc.}$$

so

$$x = -6.9°, \ 113.1°, \ 353.1°, \text{ etc.}$$

The solutions in the required range are

$$x = 113.1° \quad \text{and} \quad 353.1°.$$

Exercise 3L

1. Express each of the following in the form $R\sin(x+\alpha)$.

    (a) $12\sin x + 5\cos x$ (d) $6\sin x - 8\cos x$

    (b) $12\sin x - 5\cos x$ (e) $5\sin x - \cos x$

    (c) $\sin x + 2\cos x$ (f) $6\sin x - 3\cos x$

2. By expanding $R\cos(x-\alpha)$, find an alternative form for $A\sin x + B\cos x$, expressing R and α in terms of A and B.

3. Express each of the following in the form $R\cos(x-\alpha)$.

    (a) $5\sin x + 12\cos x$ (c) $2\sin x - \cos x$

    (b) $6\sin x - 3\cos x$ (d) $\sin x + 3\cos x$

4. By referring back to problem 1, solve the equations below in the range $-180° \le x \le 180°$.

    (a) $12\sin x + 5\cos x = 6$ (d) $6\sin x + 8\cos x = 10$

    (b) $12\sin x - 5\cos x = 0$ (e) $5\sin x - \cos x = 1$

    (c) $\sin x + 2\cos x = 1$ (f) $6\sin x - 3\cos x = 5$

4

Sequences and series

4.1   SEQUENCES

A sequence is defined as a set of numbers in a specified order, so that each number is related to the next by a rule. Each of the numbers in the sequence is called a *term*. Some examples of sequences are given below.

$$1, 8, 15, 22, 29, 36, ...$$
$$1, 4, 9, 16, 25, 36, ...$$
$$2, 1, \tfrac{1}{2}, \tfrac{1}{4}, \tfrac{1}{8}, \tfrac{1}{16}, ...$$

An *expression* can be found for each of these sequences that describes how each term is calculated. The notation u_n is used to denote the nth term of the sequence. The first term is u_1, the second u_2, etc. Each of the sequences above is now given with a rule for u_n:

$$1, 8, 15, 22, 29, 36, ...   \qquad u_n = 7n - 6$$
$$1, 4, 9, 16, 25, 36, ...   \qquad u_n = n^2$$
$$2, 1, \tfrac{1}{2}, \tfrac{1}{4}, \tfrac{1}{8}, \tfrac{1}{16}, ...   \qquad u_n = 4(\tfrac{1}{2})^n \ \text{ or } \ u_n = (\tfrac{1}{2})^{n-2}$$

The terms of the first two sequences given above increase indefinitely. The terms of the third sequence however get smaller and smaller, getting closer and closer to zero. We say that the limit of the sequence as n tends to infinity is zero. This is written as

$$\lim_{n \to \infty} \left(\tfrac{1}{2}\right)^{n-2} = 0 \ .$$

This means that as n tends to infinity (becomes larger and larger) the terms get closer and closer to zero, but never actually become zero.

DERIVE ACTIVITY 4A

It is possible to produce the terms of a sequence and find the limits of sequences using DERIVE.

(A) (i) **Author** $4(\frac{1}{2})^n$. Now **Author** the expression, VECTOR $(4(\frac{1}{2})^n, n, 10)$ and **approX**. You will obtain a vector whose elements are the terms of the sequence.

   (ii) To find a particular term of the sequence you can use the **Manage Substitute** commands. Show that the 15th term of this sequence is 1/8192.

   (iii) The terms that we have seen suggest that the limit of the sequence is zero. This can be confirmed by using the **Calculus Limit** commands, highlighting the expression for u_n and entering the point as inf. This will give the limit as n tends to ∞. **Simplifying** the expression gives 0.

(B) For each of the sequences defined below

   (i) Find the first 10 terms using DERIVE.

   (ii) Write down the limit of the sequence if it exists.

   (iii) Check your response to (ii) using the **Calculus Limit** commands.

(a) $u_n = \dfrac{1}{1+n}$ (b) $u_n = 4 + \dfrac{1}{n}$ (c) $u_n = \dfrac{n^3 + 1}{n}$

(d) $u_n = \dfrac{n^2 + 10}{n^2}$ (e) $u_n = 1.1^n$ (f) $u_n = 0.9^n$

Exercise 4A

1.   Below are listed four sequences of numbers.

     (a)   4, 7, 10, 13, 16, ...

     (b)   3, 6, 11, 18, 27, ...

     (c)   0, 3, 8, 15, 24, ...

     (d)   4, 9, 14, 19, 24, ...

     Select the rule below which defines each series.

$$u_n = n^2 - 1 \quad u_n = n^2 + 2 \quad u_n = 5n - 1 \quad u_n = 3n + 1$$

2.   Write down the first five terms of each sequence defined below and state the limit of the sequence if one exists.

     (a) $u_n = 2^n$ (b) $u_n = \left(\frac{1}{3}\right)^{n-1}$ (c) $u_n = (-2)^n$

     (d) $u_n = \left(-\frac{1}{2}\right)^n$ (e) $u_n = \dfrac{2n+1}{n}$ (f) $u_n = n + (0.1)^n$

3.   For each sequence find a rule for obtaining the *n*th term u_n. Also find the limit of each sequence if it exists.

     (a)   1, 3, 9, 27, ...

     (b)   14, 9, 4, –1, ...

     (c)   5.1, 5.01, 5.001, 5.0001, ...

     (d) $\frac{1}{3}, \frac{1}{2}, \frac{3}{5}, \frac{2}{3}, ...$

4.2   SERIES

A *series* is formed when the terms of a sequence are added together. Some examples of series are

$$1 + 4 + 9 + 16 + 25 + \dots$$
$$5 + 7 + 9 + 11 + 13 + \dots$$
$$1 + 2 + 4 + 8 + 16 + \dots$$

The symbol Σ is used to denote the sum of the terms of a series. The notation $\sum_{i=1}^{n} u_i$ is used to denote the sum of the first n terms of the series. This can be expressed as

$$\sum_{i=1}^{n} u_i = u_1 + u_2 + u_3 + \dots + u_n \ .$$

Example 4A

Write down the series described by

$$\sum_{i=1}^{6} (i^2 - 1)$$

and find the sum of this series.

Solution

The first six terms of the series must be evaluated.

$$u_1 = 1^2 - 1 = 0 \qquad\qquad u_4 = 4^2 - 1 = 15$$
$$u_2 = 2^2 - 1 = 3 \qquad\qquad u_5 = 5^2 - 1 = 24$$
$$u_3 = 3^2 - 1 = 8 \qquad\qquad u_6 = 6^2 - 1 = 35$$

Now these terms can be put into the series

$$\sum_{i-1}^{6} (i^2 - 1) = 0 + 3 + 8 + 15 + 24 + 35 = 85 \ .$$

So the sum of the series is 85.

DERIVE ACTIVITY 4B

It is possible to find the sum of a series using DERIVE.

(A) (i)   **Author** the expression for u_i the ith term of the series. As an example
          use $u_i = 3^i$.

     (ii) With the expression highlighted use the **Calculus** and **Sum** commands,
          with lower limit 1 and upper limit 7 to obtain

$$\sum_{i=1}^{7} 3^i \; .$$

     (iii) Simplify this expression to obtain the value of the sum.

(B) Use DERIVE to evaluate the following sums

   (a) $\displaystyle\sum_{i=1}^{10} i$ (b) $\displaystyle\sum_{i=1}^{20} (i)^2$ (c) $\displaystyle\sum_{i=1}^{50} \frac{1}{i}$.

(C) Some series will converge to a limit when they are summed, but others will
     increase indefinitely or diverge. By using **Calculus Sum** with an upper limit of
     ∞ (inf), investigate the problems below.

   (i)   Consider $\displaystyle\sum_{i=1}^{\infty} \frac{1}{a^i}$ for different values of a. Try $a = 1, 2, 3$. For what

         values of a does the series converge to a limit?

   (ii) Consider $\displaystyle\sum_{i=1}^{\infty} \frac{1}{i^n}$ for different values of n. Try $n = 1, 2, 3$. For what

         values of n does the series converge to a limit?

The convergence of infinite series is a fascinating subject which is not studied in this elementary text. Whether an infinite series will converge may not always be obvious from the sequence forming the series. For example, the series $\sum_{i=1}^{\infty} \frac{1}{i}$ does not converge even though $\lim_{n \to \infty} u_n = 0$. There are many mathematical tests for convergence. A simple test is that the series $\sum_{i=1}^{\infty} u_i$ will converge if $\lim_{n \to \infty} \left| \frac{u_{n+1}}{u_n} \right| < 1$. This is called the *ratio test*. Note that for $\sum_{i=1}^{\infty} \frac{1}{i}$, we have $\lim_{n \to \infty} \frac{u_{n+1}}{u_n} = 1$.

Exercise 4B

1. Write each of the series below in full and find their sum.

    (a) $\sum_{i=1}^{4} (i+1)^2$ (b) $\sum_{i=1}^{6} \frac{1}{i}$ (c) $\sum_{i=1}^{4} (4i-2)$

    (d) $\sum_{i=1}^{5} 2i$ (e) $\sum_{i=1}^{3} \frac{1}{i^2}$ (f) $\sum_{i=1}^{4} (i^3 - 3)$

2. Write each of the series given below using sigma notation.

    (a) $16 + 11 + 6 + 1 - 4$ (b) $2 + 6 + 18 + 54$

    (c) $\frac{1}{2} + 2 + 4\frac{1}{2} + 8 + 12\frac{1}{2} + 18$ (d) $1 - 2 + 4 - 8 + 16$

3. Apply the ratio test to each of the series in problem 1 to suggest which infinite series would converge to a limit. For those which do converge use DERIVE to find the limit.

4.3   ARITHMETIC PROGRESSIONS

An *arithmetic progression* (AP) is a sequence where each term is obtained from the previous term by adding or subtracting the same number each time. The example below is a sequence where 3 is added each time to obtain the next term.

2, 5, 8, 11, 14, 17.

The number that is added each time is known as the *common difference* and usually represented by the letter *d*. The first term is usually represented by the letter *a*.

In general an arithmetic progression will have terms given by

$a, a + d, a + 2d, a + 3d, \dots$

So the *n*th term u_n will be given by

$u_n = a + (n-1)d$.

Example 4B

An AP has first term 5 and common difference 4.

(a)   Write down the first 6 terms of the AP.

(b)   Find an expression for u_n, the *n*th term.

(c)   Find the 50th term using the answer to (b).

Solution

(a)   The AP will begin with 5 and subsequent terms can be obtained by adding 4 to the previous term to give

5, 9, 13, 17, 21, 25.

(b)   The *n*th term is given by

$u_n = a + (n-1)d$.

Here $a = 5$ and $d = 4$ so,

$u_n = 5 + (n-1)4 = 5 + 4n - 4 = 1 + 4n$.

(c)   The 50th term is given by u_{50}.

$$u_{50} = 1 + 4 \times 50 = 201.$$

Example 4C

The sequence

2, 4, 6, 8, 10, 12, 14, 16, 18, 20

is an AP. Find the sum of its terms.

Solution

The sum of its terms, S, is given by the series

$$S = 2 + 4 + 6 + \ldots + 16 + 18 + 20$$

Consider the series written in reverse order.

$$S = 20 + 18 + 16 + \ldots + 6 + 4 + 2$$

Now adding the two series together gives

$$2S = 22 + 22 + 22 + \ldots + 22 + 22 + 22 .$$

Noting that the series has 10 terms gives

$$2S = 10 \times 22$$

so

$$S = \frac{10 \times 22}{2} = 110 .$$

The Sum of an Arithmetic Progression

Using the approach of Example 4C we can find a general formula for the sum of an AP. The sum of the first n terms of an AP, S_n, is given by

$$S_n = a + (a+d) + \cdots + (a + (n-2)d) + (a + (n-1)d) .$$

Writing the series in reverse order

$$S_n = (a + (n-1)d) + (a + (n-2)d) + \cdots + (a+d) + a$$

and adding the two series gives

$$2S_n = (a + a + (n-1)d) + (a + d + a + (n-2)d) + \cdots + (a + d + a + (n-2)d) + (a + a + (n-1)d)$$
$$2S_n = (2a + (n-1)d) + (2a + (n-1)d) + \cdots + (2a + (n-1)d) + (2a + (n-1)d)$$

There are n identical terms on the right hand side so that

$$S_n = \frac{n(2a + (n-1)d)}{2}$$

The sum of an AP

$$u_n = a + (n-1)d$$

is

$$S_n = \frac{n(2a + (n-1)d)}{2}$$

Example 4D

Find the sum of the arithmetic progression

$$1 + 8 + 15 + ... + 260 .$$

Solution

This is an AP with $a = 1$ and $d = 7$. To find n use

$$u_n = a + (n - 1)d$$

applied to the last term. This gives

$$260 = 1 + 7(n-1)$$

$$260 = 7n - 6$$

$$n = \frac{260 + 6}{7}$$

$$= 38 .$$

Now the sum can be found

$$S_{38} = \frac{38(2 \times 1 + (38 - 1) \times 7)}{2}$$

$$= 4959 .$$

Exercise 4C

1.   Find the 8th term of each of the APs given below.

   (a)   1, 10, 19, 28, ...

   (b)   4, 3, 2, 1, ...

   (c)   5, 7, 9, 11, ...

   Also write down an expression for the nth term of each AP.

2.   For each AP below find the common difference and the number of terms.

   (a)   4, 7, 10, ..., 85

   (b)   −1, 3, 7, ..., 67

   (c)   51, 44, 37, ..., −54

3.   Find the sum of each of the following series.

   (a) $1 + 4 + 7 + 10 + ... + 70$

   (b) $6 + 4 + 2 + ... - 18$

   (c) $10 + 10.1 + 10.2 + ... + 20$

4.   Find the sum of the first 10 terms of the APs that begin

   (a)   1, 3, 5, ...

   (b)   7, 10, 13, ...

5.   Find the sum of the first 20 even numbers.

6.   In an AP the fifth term is 16 and the second term 7. Find the first term.

7.   The fifth term of an AP is 22 and the sum of the first 10 terms is 240. Find the first term and the common difference.

4.4   GEOMETRIC PROGRESSIONS

In a *geometric progression* (GP) each term is obtained from the previous term by multiplying by the same number each time. For example the sequence below is obtained by multiplying the previous term by 2.

3, 6, 12, 24, 48, 96, ...

The number that is used to multiply the terms of the GP is known as the *common ratio*, usually referred to as r. In the above example the common ratio is 2.

A GP with first term a and common ratio r would be

$$a, ar, ar^2, ar^3, ar^4, ...$$

The nth term of such a sequence would be given by $u_n = ar^{n-1}$.

Example 4E

The first two terms of a GP are 16 and 22.4.

(a)   Find the common ratio.

(b)   Give an expression for the nth term.

(c)   Find the 4th term of the GP.

Solution

(a)   To find the common ratio note that the first term is a and the second ar, so

$$r = \frac{ar}{a} = \frac{u_2}{u_1} .$$

In fact the ratio of any 2 successive terms could be used. In this case

$$r = \frac{22.4}{16} = 1.4 .$$

(b)   The nth term is given by

$$u_n = ar^{n-1}$$

so here

$$u_n = 16 \times 1.4^{n-1} .$$

(c)   The 4th term is given by

$$u_4 = 16 \times 1.4^3$$
$$= 16 \times 2.744$$
$$= 43.904 .$$

The Sum of a GP

We now look for a general formula for the sum of a GP with n terms. Using S_n to denote this sum,

$$S_n = a + ar + ar^2 + \ldots + ar^{n-1} .$$

This expression is now multiplied by r to give

$$rS_n = ar + ar^2 + ar^3 + \ldots + ar^{n-1} + ar^n .$$

Subtracting the second of these expressions from the first gives

$$S_n - rS_n = a - ar^n$$

or

$$S_n(1-r) = a(1-r^n)$$

so that

$$S_n = \frac{a(1-r^n)}{(1-r)} .$$

Example 4F

Find the sum of the first 20 terms of the GP

$$4, 6.8, 11.56, 19.652, \ldots$$

Solution

Here $a = 4$. The common ratio r is given by

$$r = \frac{6.8}{4} = 1.7 .$$

So the sum of the first 20 terms, S_{20}, is given by

$$S_{20} = 4\frac{(1-1.7^{20})}{(1-1.7)} = 232236 \qquad \text{to 6 significant figures.}$$

The Sum of an Infinite Geometric Progression

To find the sum of a GP with an infinite number of terms we consider the limit of S_n as $n \to \infty$.

$$\lim_{n \to \infty} a\left(\frac{1-r^n}{1-r}\right)$$

As only the r^n part of this expression depends on n, it is the behaviour of this that is important. If $-1 < r < 1$ then as $n \to \infty$, $r^n \to 0$, so

$$\lim_{n \to \infty} a\left(\frac{1-r^n}{1-r}\right) = a\left(\frac{1-0}{1-r}\right)$$

$$= \frac{a}{1-r}.$$

If $r > 1$ then $r^n \to \infty$ as $n \to \infty$.

If $r < -1$ then r^n will alternate between positive and negative values, however $|r^n|$ will also become bigger and bigger.

So the sum of an infinite GP will only converge if $-1 < r < 1$.

Example 4G

The first four terms of an infinite GP are given by

$$100, \ 99, \ 98.01, \ 97.0299.$$

Find its sum.

Solution

Here $a = 100$ and $r = 0.99$.

The sum S is given by

$$S = \frac{a}{1-r} = \frac{100}{1-0.99} = \frac{100}{0.01} = 10000.$$

Exercise 4D

1.   For each of the geometric progressions below, find

      (i)   the common ratio,
      (ii)   the 10th term,
      (iii)   the sum of the first 8 terms.

      (a)   4, 4.8, 5.76, ...
      (b)   1.3, 1.04, 0.832, ...
      (c)   −1.7, 0.85, −0.425, ...
      (d)   200, 320, 512, ...

2.   For each GP below find the sum to infinity, if it exists.

      (a)   40, 16, 6.4, ...

      (b)   0.5, 0.6, 0.72, ...

      (c)   1, 0.9, 0.81, 0.729

      (d)   −1, 0.9, −0.81, 0.729, ...

3.   A ball is dropped from a height of 1m. Each time it bounces it rebounds to 2/5
      of the height from which it was released. How far does it travel before it stops?

4.   The second and third terms of a geometric progression are 12 and $6(c+1)$
      respectively. Find c if the sum of the first three terms of the GP is 38.

5.   Show that if you invest £1000 in a building society at an annual interest rate of
      r% then the amount after n years forms the geometric progression

$$u_n = 1000\left(1 + \frac{r}{100}\right)^n .$$

      [Assume that you do not remove any money!].

4.5   THE BINOMIAL THEOREM

In this section we deal with expanding expressions like $(2+x)^{12}$ or $(5-x)^4$. This is then developed to enable series expansions to be made for expressions like $(1-x)^{1/2}$ or $(1+x^2)^{-2}$.

Expanding $(a+b)^n$: n a Positive Integer

DERIVE ACTIVITY 4C

(A)   (i)   In DERIVE **Author** and **Expand** the following expressions:

$$(1+x)^2,\ (1+x)^3,\ (1+x)^4,\ (1+x)^5.$$

What do you notice about the powers of x that you obtain?

What do you notice about the coefficients of the terms you obtain?

(ii)   If you expand $(1+x)^6$ what would you expect the first 2 and the last 2 terms to be?

Try it to check your prediction.

(iii)   The number pattern below is known as *Pascal's Triangle*.

$$
\begin{array}{ccccccccccc}
 & & & & 1 & & 1 & & & & \\
 & & & 1 & & 2 & & 1 & & & \\
 & & 1 & & 3 & & 3 & & 1 & & \\
 & 1 & & 4 & & 6 & & 4 & & 1 & \\
1 & & 5 & & 10 & & 10 & & 5 & & 1 \\
\end{array}
$$

How do the coefficients of your expressions relate to the triangle?

What would you expect the next two rows of the triangle to be?

Now **Author** and **Expand** $(1+x)^7$ to check your prediction.

(B)   (i)   **Author** and **Expand** $(a+b)^1,\ (a+b)^2,\ (a+b)^3,\ (a+b)^4$.
Do the coefficients still relate to Pascal's Triangle?
What happens to the powers of a and b?

(ii)   Using Pascal's Triangle write down the expansion of $(a+b)^5$. Check your prediction on DERIVE.

(C) (i) Pascal's Triangle provides a good approach for small values of n. Now consider $(a+b)^{12}$. DERIVE can expand this easily, try it and see.

     (ii) Now **Author** and **Simplify** COMB(12,4). Which coefficient does it give you?

     (iii) Repeat (ii) for COMB(12,5) and COMB(12,1).

     (iv) **Author** and **Simplify** VECTOR(COMB(12,n),n,0,12). Check that this gives you all the coefficients of expansion.

     (v) Use the VECTOR and COMB commands to find the coefficients of $(a+b)^{10}$. Then write down the expansion of $(a+b)^{10}$. Check your work on DERIVE by expanding $(a+b)^{10}$.

     (vi) **Author** $\dfrac{n!}{r!(n-r)!}$. Using **Manage**, **Substitute** replace n by 10 and r by 0,1,2,...,10. How do you think DERIVE calculated COMB(n,r)?

     Normally COMB(n,r) is written ${}^nC_r$ or $\binom{n}{r}$. In this book ${}^nC_r$ will be used.

     If you have not met the ! symbol (read as factorial) before, try **Author** and **Simplify** 1!, 2!, 3!, 4! and 5!.

     (vii) **Author** and **Simplify** COMB(n,0), COMB(n,1), COMB(n,2), COMB(n,3). What do you notice?

(D) (i) Highlight your expansion of $(a+b)^4$. Use the **Manage** and **Substitute** commands to replace a by 2 and b by $3x$. What happens? Check that what you have obtained is the expansion of $(2+3x)^4$.

     (ii) By using your expansion of $(a+b)^5$, find $(3+5x)^5$.

     (iii) Repeat (ii) for $(x+\frac{1}{2})^5$, $(1+3x)^5$ and $(6-5x)^5$.

Summary

The expansion of $(a+b)^n$ where n is a positive integer is given by

$$(a+b)^n = {}^nC_0 a^n + {}^nC_1 a^{n-1}b + {}^nC_2 a^{n-2}b^2 + \cdots + {}^nC_n b^n$$

where ${}^nC_r = \dfrac{n!}{r!(n-r)!}$. This is called *the Binomial Theorem.*

Example 4H

Expand each of the expressions below using the Binomial Theorem.

(a) $(1+x)^6$

(b) $(2+3x)^4$

Solution

(a)   Taking $a = 1$ and $b = x$ gives

$$(1+x)^6 = {}^6C_0 1^6 x^0 + {}^6C_1 1^5 x^1 + {}^6C_2 1^4 x^2 + {}^6C_3 1^3 x^3$$
$$+ {}^6C_4 1^2 x^4 + {}^6C_5 1^1 x^5 + {}^6C_6 1^0 x^6$$
$$= 1 + 6x + 15x^2 + 20x^3 + 15x^4 + 6x^5 + x^6$$

(b)   Taking $a = 2$ and $b = 3x$ gives

$$(2+3x)^4 = {}^4C_0 2^4 (3x)^0 + {}^4C_1 2^3 (3x)^1 + {}^4C_2 2^2 (3x)^2 + {}^4C_3 2^1 (3x)^3$$
$$+ {}^4C_4 2^0 (3x)^4$$
$$= 1 \times 16 \times 1 + 4 \times 8 \times 3x + 6 \times 4 \times 9x^2 + 4 \times 2 \times 27x^3 + 1 \times 1 \times 81x^4$$
$$= 16 + 96x + 216x^2 + 216x^3 + 81x^4 .$$

Example 4I

Find the first four terms of the expansion of $(1+x)^n$.

Solution

Note first that

$$^nC_0 = \frac{n!}{0!(n-0)!} = 1$$

$$^nC_1 = \frac{n!}{1!(n-1)!} = n$$

$$^nC_2 = \frac{n!}{2!(n-2)!} = \frac{n(n-1)}{2!}$$

$$^nC_3 = \frac{n!}{3!(n-3)!} = \frac{n(n-1)(n-2)}{3!} .$$

The first four terms will be given by

$$(1+x)^n = {}^nC_0 + {}^nC_1 x + {}^nC_2 x^2 + {}^nC_3 x^3$$
$$= 1 + nx + \frac{n(n-1)}{2!} x^2 + \frac{n(n-1)(n-2)}{3!} x^3$$

Note that the pattern of terms will continue with the rth term given by

$$\frac{n(n-1)(n-2)\cdots(n-r+1)}{r!} x^r .$$

Also note that as n is a positive integer, then the series will terminate when $n = r$.

Exercise 4E

1.   Expand each of the following.

    (a) $(1+x)^4$ (b) $(2+x)^3$

    (c) $(1-x)^5$ (d) $(5+3x)^4$

    (e) $(5-2x)^3$ (f) $(x-y)^5$

    (g) $\left(u+\dfrac{1}{u}\right)^4$ (h) $\left(2x+\dfrac{1}{x}\right)^4$

2.   Find the coefficient of x^3 in each of the expansions below.

    (a) $(2+5x)^{10}$ (b) $(3-2x)^4$

    (c) $(8-7x)^3$ (d) $(3-2x^3)^2$

3.   Find the coefficient of x^7 in each of the expansions below.

    (a) $(1+x)^{13}$ (b) $(1-2x)^{11}$

    (c) $(3+0.5x)^{13}$ (d) $(1.3-0.1x)^9$

    (e) $(a+bx)^{15}$ (f) $(a-bx)^{14}$

4.   Write each of the following in the form $(a+bx)^n$.

    (a) $1+6x+15x^2+20x^3+15x^4+6x^5+x^6$

    (b) $0.0625+0.5x+1.5x^2+2x^3+x^4$

    (c) $1-2x+1.5x^2-0.5x^3+0.0625x^4$

The Binomial Theorem for Any Index

DERIVE ACTIVITY 4D

The previous section has shown how the Binomial Theorem can be applied to expand $(1+x)^n$, where n is a positive integer. Now we will investigate what happens for other values of n.

(A) (i)   **Author** the expression

$$1 + nx + \frac{n(n-1)x^2}{2} + \frac{n(n-1)(n-2)x^3}{6} \; .$$

(ii) Use **Manage** and **Substitute** to replace n by $\frac{1}{2}$. **Simplify** and **Plot** the expression you obtain.

(iii) Now **Author** and **Plot** $(1+x)^{\frac{1}{2}}$ or $\sqrt{(1+x)}$. For what range of values does your series give a good approximation to the function.

(B) Extend your binomial series by adding the terms

$$\frac{n(n-1)(n-2)(n-3)x^4}{4!} + \frac{n(n-1)(n-2)(n-3)(n-4)x^5}{5!} \; .$$

Investigate how this improves the quality of the approximation.
You may wish to add further terms.

(C) Substitute each value of n given below into the expression you authored in (A) (i). **Simplify** and **Plot** it and then compare it with the plot of the function $(1+x)^n$. Comment on the range of values for which the series gives a good approximation.

(i) $n = -1$ (ii) $n = \frac{3}{2}$

(iii) $n = \frac{1}{3}$ (iv) $n = -2$

Can you see a pattern emerging? What can you deduce about the binomial theorem for negative and fractional powers of n?

Summary

You have seen that $(1+x)^n$ gives a finite series when n is a positive integer. The notation ^nC_r would have no meaning when n is not a positive integer, but it is possible to produce a series expansion of $(1+x)^n$ for other values of n by using

$$(1+x)^n = 1 + nx + \frac{n(n-1)x^2}{2!} + \frac{n(n-1)(n-2)}{3!}x^3 + \cdots$$

Here neither $n!$ or ^nC_r are used. You will have seen in DERIVE Activity 4D that this type of series gives a good approximation to $(1+x)^n$ for small values of x.

It is important to note that for fractional or negative values of n the series will never terminate and will have an infinite number of terms. The series will only converge if $-1 < x < 1$ or $|x| < 1$. The more terms that you take in a series, the better approximation it will give to $(1+x)^n$.

Example 4J

Find the first four terms of the expansion of $(1+x)^n$ for (a) $n = -1$ and (b) $n = \frac{1}{2}$.

Solution

The terms required are given below

$$(1+x)^n = 1 + nx + \frac{n(n-1)x^2}{2!} + \frac{n(n-1)(n-2)}{3!}x^3 + \cdots$$

(a)   For $n = -1$

$$(1+x)^n = 1 + (-1)x + \frac{(-1)(-1-1)x^2}{2!} + \frac{(-1)(-1-1)(-1-2)x^3}{3!} + \cdots$$

$$= 1 - x + x^2 - x^3 + \cdots$$

(b)   For $n = \frac{1}{2}$

$$(1+x)^{\frac{1}{2}} = 1 + (\tfrac{1}{2})x + \frac{(\tfrac{1}{2})(\tfrac{1}{2}-1)x^2}{2!} + \frac{(\tfrac{1}{2})(-\tfrac{1}{2}-1)(-\tfrac{1}{2}-2)x^3}{3!} + \cdots$$

$$= 1 + \frac{x}{2} - \frac{x^2}{8} + \frac{x^3}{16} + \cdots$$

Example 4K

Expand $(1+3x)^{-2}$ using the binomial theorem and state the range of values for which it converges.

Solution

Using the expansion,

$$(1+x)^n = 1 + nx + \frac{n(n-1)x^2}{2!} + \cdots$$

with $n = -2$ gives

$$(1+3x)^{-2} = 1 + (-2)(3x) + \frac{(-2)(-2-1)(3x)^2}{2!} + \frac{(-2)(-2-1)(-2-2)(3x)^3}{3!} + \cdots$$

$$= 1 - 6x + 27x^2 - 108x^3 + \cdots$$

For the series to converge $-1 < 3x < 1$ or $-\frac{1}{3} < x < \frac{1}{3}$.

Exercise 4F

1.   Find the first four terms of the binomial expansion for each expression.

   (a) $(1+x)^{-2}$ (b) $(1+x)^{-3}$

   (c) $(1+x)^{3/2}$ (d) $(1+x)^{-1/2}$

2. State the values of x for which a binomial expansion of the following expressions would converge.

    (a) $\dfrac{1}{1-2x}$ (b) $\sqrt{1+\frac{1}{3}x}$

    (c) $\dfrac{1}{(1+4x)^2}$ (d) $\dfrac{1}{\sqrt{1-5x}}$

3. Find the first four terms of the binomial expansions of each expression, stating the range of values of x for which it converges.

    (a) $(1+3x)^{\frac{1}{2}}$ (b) $(1-2x)^{-2}$

    (c) $\dfrac{1}{1+4x}$ (d) $\dfrac{1}{\sqrt{1+2x}}$

4. Use the expansion of $(1+x)^{\frac{1}{2}}$ obtained in Example 4J to evaluate $\sqrt{0.98}$ and $\sqrt{1.2}$ by substituting suitable values of x.

5. (a) Show that

$$\frac{1}{(3+x)^2} = \frac{1}{x^2\left(\dfrac{3}{x}+1\right)^2} .$$

    (b) Find the first four non-zero terms of the expansion of $\left(\dfrac{3}{x}+1\right)^{-2}$.

    (c) State the range of values of x for which this expansion converges.

    (d) Use your answers to find a series expansion for $(3+x)^{-2}$.

6. Find expansions for each of the expressions below.

    (a) $\dfrac{1}{2+x}$ (b) $\dfrac{1}{4+2x}$

    (c) $\dfrac{1}{(5+x)^2}$ (d) $\sqrt{2+x}$

5

Simple numerical methods for solving equations

5.1   DECIMAL SEARCH

DERIVE ACTIVITY 5A

(A)   Suppose that you have to find the solutions (or roots) of the equation
$3x^3 + x^2 - 5x - 1 = 0$.

    (i)   In DERIVE **Author** and **Plot** $3x^3 + x^2 - 5x - 1$ as shown in Figure 5.1.
From the plot it should be clear that one solution lies between -2 and -1,
another between -1 and 0 and the third solution between 1 and 2.

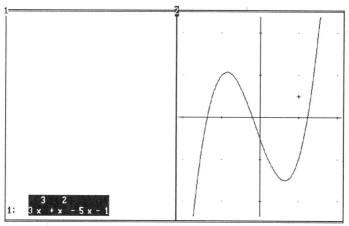

Figure 5.1

(ii)   We will concentrate on finding the positive solution. Return to the Algebra window and **Manage Substitute** the values 1 and 2 into the equation of the curve and **approX**. Note that one gives a positive value and the other a negative value. This shows that the curve must cross the axis between these two points.

(iii)   Now **Manage, Substitute** 1.2 and 1.3 into the equation and **approX**. What can you say about the solution?

(iv)   Now **Manage, Substitute** 1.23 and 1.24 and **approX**. What can you conclude about the solution?

(v)   Now try 1.234, 1.235 and 1.2345.

(vi)   Give the solution correct to 3 decimal places.

(vii)   By trying further values of your own find the solution correct to 4 decimal places.

(B)   Use an approach similar to (A), but substituting values of your own choice to find the other two roots of the equation.

(C)   Use an approach similar to that described in (A) to find the solutions of

(i) $x^3 + 3x^2 - 1$ (positive solution only)

(ii) $e^x - x^2 = 0$

(iii) $x - \cos x = 0.$

Summary

The DERIVE activity has shown you how to find the solutions of the equation $f(x) = 0$, by trying a range of values of x. If $f(x)$ changes from being positive to negative between any two values, then a solution must lie between these two values. Further intermediate values of x can then be tried until the solution is obtained to the required degree of accuracy. This is called the *decimal search* method.

Example 5A

For each of the functions f find the solution of $f(x) = 0$ that lies between 1 and 2.

(a) $f(x) = \dfrac{x}{2} - \dfrac{1}{x}$ (b) $f(x) = x^3 - 7$

Solution

(a)   Table 5.1 shows the process for $f(x) = \dfrac{x}{2} - \dfrac{1}{x}$.

From $f(1) < 0$ and $f(2) > 0$ we deduce that the solution lies between $x = 1$ and $x = 2$.

Since $f(1.5) > 0$ we deduce that the solution lies between $x = 1$ (for which $f(1) < 0$) and $x = 1.5$.

And so on.

The third column (comments on solution) summarises this information.

Table 5.1

Trial Value of x	Value of $f(x)$ (to 4 dp)	Comments on solution
1	−0.5	-
2	0.5	$1 \quad < x < 2$
1.5	0.0833	$1 \quad < x < 1.5$
1.4	−0.0143	$1.4 < x < 1.5$
1.45	0.0353	$1.4 < x < 1.45$
1.42	0.0058	$1.4 < x < 1.42$
1.41	−0.0042	$1.41 < x < 1.42$
1.415	0.0008	$1.41 < x < 1.415$

The solution can now be quoted as 1.41 correct to 2 decimal places. Further work would lead to a more accurate solution.

(b)   Table 5.2 shows the process for $f(x) = x^3 - 7$.

Table 5.2

Trial Value of x	Value of $f(x)$	Comments on solution
1	-6	-
2	1	$1 \quad < x < 2$
1.8	-1.168	$1.8 \quad < x < 2$
1.9	-0.141	$1.9 \quad < x < 2$
1.95	0.41488	$1.9 \quad < x < 1.95$
1.92	0.07789	$1.9 \quad < x < 1.92$
1.91	-0.03213	$1.91 \quad < x < 1.92$
1.912	-0.01022	$1.912 \quad < x < 1.92$
1.913	0.00076	$1.912 \quad < x < 1.913$
1.9125	-0.00473	$1.9125 < x < 1.913$

Now the solution can be given as 1.913 correct to 3 decimal places.

Exercise 5A

1.   Show that the equation $x^4 - 5$ has a solution between 1 and 2. Find this solution correct to 2 decimal places.

2.   Show that the equation $x^3 - x + 10 = 0$ has a solution between -2 and -3. Find this solution correct to 2 decimal places.

3.   Show that the equation $\sin(x) - x + 1 = 0$ has a solution between 1 and 2. Find this solution correct to 3 decimal places working in radians.

4.   Sketch a graph of $y = e^x + x$. Use this to help you find the solution of $e^x + x = 0$.

5.   (a)   Find the solution of the equation $x^3 - 10 = 0$.

      (b)   Find the cube root of 20.

5.2   FORMULA ITERATION

While a decimal search will tend to the solutions of an equation there are more systematic and efficient approaches that can be taken. An *iterative formula* is simply a formula that will produce a sequence of numbers that converge to a particular value or may diverge. These formulae are defined in the form

$$x_{n+1} = f(x_n) \,.$$

This defines x_{n+1} (the next number in the sequence) in terms of the previous number x_n.

Example 5B

Find the first 5 terms of the sequence defined by

$$x_{n+1} = \frac{1}{2}\left(x_n + \frac{2}{x_n}\right)$$

if $x_1 = 1$.

Solution

We have $x_1 = 1$. The next term x_2 can be obtained by substituting x_1 into the iterative formula to give

$$x_2 = \frac{1}{2}\left(x_1 + \frac{2}{x_1}\right) = \frac{1}{2}\left(1 + \frac{2}{1}\right) = 1.5 \,.$$

Similarly for x_3, x_4 and x_5

$$x_3 = \frac{1}{2}\left(x_2 + \frac{2}{x_2}\right) = \frac{1}{2}\left(1.5 + \frac{2}{1.5}\right) = 1.417 \quad (\text{to 3 dp}).$$

$$x_4 = \frac{1}{2}\left(1.417 + \frac{2}{1.417}\right) = 1.414 \quad (\text{to 3 dp}).$$

$$x_5 = \frac{1}{2}\left(1.414 + \frac{2}{1.414}\right) = 1.414 \quad (\text{to 3 dp}).$$

The iterative formula in Example 5B defines a sequence that converges to a value of 1.414 correct to 3 decimal places.

Consider the general problem of solving $f(x) = 0$. It is always possible to formulate an iterative formula by rearranging the equation to be solved into the form $x = g(x)$. For example consider the equation

$$x^3 + x - 3 = 0 .$$

One possible rearrangement is

$$x = 3 - x^3$$

and would lead to the iterative formula

$$x_{n+1} = 3 - x_n^3 .$$

Another possibility is

$$x^3 = 3 - x$$

$$x = \frac{3 - x}{x^2}$$

which would lead to the iterative formula

$$x_{n+1} = \frac{3 - x_n}{x_n^2} .$$

Another possibility is

$$x^3 = 3 - x$$

$$x^2 = \frac{3 - x}{x}$$

$$x = \sqrt{\frac{3 - x}{x}}$$

which would give an iterative formula

$$x_{n+1} = \sqrt{\frac{3 - x_n}{x_n}} .$$

Which of these iterative formulas leads to the solution? The DERIVE plot of $x^3 + x - 3$, in figure 5.2, shows that there is a solution close to $x = 1$.

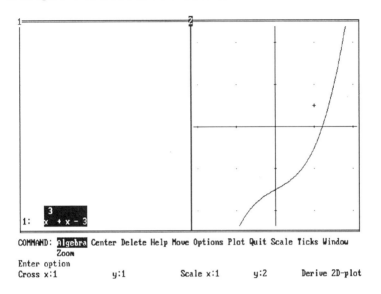

Figure 5.2

Table 5.3 below gives a comparison of each of the 3 formulae all starting with $x_1 = 1$.

Table 5.3

	$x_{n+1} = 3 - x_n^3$	$x_{n+1} = \dfrac{3 - x_n}{x_n^2}$	$x_{n+1} = \sqrt{\dfrac{3 - x_n}{x_n}}$
x_1	1	1	1
x_2	2	2	1.41421
x_3	-5	0.25	1.05892
x_4	128	44	1.35391
x_5	-2097149	-0.0212	1.10263

Here the first two formulae behave in a very strange way, but the final version seems that it may converge. After a number of further iterations (or steps) it converges to 1.21341. So when using formula iteration it is important to be careful as not all iterative formulae will converge.

DERIVE ACTIVITY 5B

This activity will introduce you to the ITERATES command that can help you to rapidly apply iterative formulae.

(A)   (i)   **Author** and **approX** the expression ITERATES$\left(\frac{1}{2}\left(x + \frac{5}{x} \right), x, 2 \right)$. You should see that this sequence converges rapidly.

(ii)   Now try the iterative formula $x_{n+1} = \sqrt{\left(\frac{3 - x_n}{x_n} \right)}$ by authoring and

approximating ITERATES$\left(\sqrt{\frac{3 - x}{x}}, x, 1 \right)$. You should see that this converges slowly.

(B)   This section considers finding the solutions of $x^2 + x - 10 = 0$ using iterative formulae.

(i)   **Author** and **Plot** $x^2 + x - 10$, and verify that one solution is close to -4 and the other is close to 3.

(ii)   By hand verify that each of the following iterative formulae can be formulated from the equation $x^2 + x - 10 = 0$.

(a) $x_{n+1} = \sqrt{10 - x_n}$ (c) $x_{n+1} = \frac{10 - x_n}{x_n}$

(b) $x_{n+1} = 10 - x_n^2$

(iii)   **Author** and **approX** the expression ITERATES$(\sqrt{10 - x}, x, 3, 10)$, which will produce the first 10 iterations starting with $x = 3$. This form of the ITERATES command is useful for formulae that may diverge as it prevents DERIVE doing endless iterations. Now try using -4 as a starting point. What do you notice?

(iv)   Repeat (iii) for each of the other iterative formulae.

(C)   Use the plot facility and the ITERATES command together with a suitable rearrangement and iterative formula to find solutions of

(i) $x^2 - 4x + 1 = 0$

(ii) $x^3 + 2x - 5 = 0$

(iii) $x^3 - 3x^2 + 4x - 2 = 0$

Exercise 5B

1.   Use the iterative formula

$$x_{n+1} = 7 - \frac{5}{x_n}$$

starting with $x_1 = 6$ to find a solution of $x^2 - 7x + 5 = 0$ correct to 4 decimal places. Can you use this iterative formula to find the root close to $x = 1$?

2.   Use the iterative formula

$$x_{n+1} = \frac{x_n^2 + 5}{7}$$

starting with $x_1 = 0$ to find a solution of $x^2 - 7x + 5 = 0$ correct to 4 decimal places. Can you use this iterative formula to find the root close to $x = 6$?

3.   (a)   Show that

$$x_{n+1} = \sqrt{\frac{30}{x_n}} \quad \text{and} \quad x_{n+1} = \frac{2x_n}{3} + \frac{10}{x_n^2}$$

are two iterative formula that could be derived from the equation $x^3 - 30 = 0$.

(b)   Using $x_1 = 3$ find x_5 for both methods.

(c)   Comment on your results and find $\sqrt[3]{30}$ correct to 4 decimal places.

4.   The iterative sequence

$$x_{n+1} = \frac{1}{2}\left(x_n + \frac{a}{x_n}\right)$$

is well known for finding solutions of the equation $x^2 = a$.

(a)   Use the formula to solve $x^2 = 8$ and $x^2 = 10$.

(b)   Show that $x = \frac{1}{2}\left(x + \frac{a}{x}\right)$ can be rearranged to give $x^2 = a$.

(c)   For what equation would the iterative formula

$$x_{n+1} = \frac{1}{3}\left(x_n + \frac{a}{x_n}\right)$$

give the solution. Use it with $a = 10$ and $x_1 = 2$ to verify your prediction.

5.   (a)   Sketch a graph of $y = x^2 - \sin(x)$ and state the value of the smallest root of $x^2 - \sin(x) = 0$.

     (b)   Find two rearrangements of the equation $x^2 - \sin(x) = 0$ and test them to see if they converge.

     (c)   Find the other root of the equation correct to 3 decimal places.

6.   Use suitable rearrangements to develop iterative formulae to solve the equations below. Find each solution correct to 2 decimal places.

     (a) $x^3 + x - 7 = 0$

     (b) $x - \cos x = 0$

     (c) $e^x - x - 2 = 0$

6

Differentiation

6.1   INTRODUCTION

In many applications of mathematics we are interested in how a quantity is changing. For example, in waiting for a cup of coffee to cool the *rate of change* of its temperature is important; the state of the economy might be measured by the *rate of change* of the money supply; the increase in the number of species of an animal could be measured by the *rate of change* of the population.

In each of these examples we need to define a function to model the situation and then find how the function is changing. In Chapter 2 you investigated the rates of change of exponential functions $f(x) = a^x$ for different values of a by drawing tangents to the graphs of the functions and finding their slopes. For example we found that when $a = 3$ the tangent slopes were larger than the function values and when $a = 2$ the tangent slopes were less than the function values. In this way you discovered the "special exponential" function e^x whose tangent slope is equal to the function value at each point.

In this chapter we link these two ideas together by defining the rate of change of a function to be the slope of the tangent to the graph of the function and calling it the *derivative* of the function. *Differentiation* is the name given to the process of finding the derivative of a function. The rules of differentiation provide an algebraic method of finding rates of change instead of the graphical approach of Chapter 2.

6.2   DIFFERENTIATION OF POLYNOMIALS

DERIVE ACTIVITY 6A

The aim of this investigation is to introduce DERIVE's commands for finding derivatives and to investigate the derivative of powers of x.

(A)   (i)   (1) **Author** x. Select **Calculus** and **Differentiate**.
Confirm variable:x and order:1.

On the screen you should see the symbols:

$$\frac{d}{dx}(x)$$

which mean, "Differentiate x once with respect to the variable x".

Now **Simplify** the expression. The result should be 1.

Repeat the process for:

(ii) x^2 (iii) x^3 (iv) x^4 (v) x^5 (vi) x^6 (vii) x^7.

Keep a record of your work.

What pattern can you see emerging?

(B)   Try to differentiate each expression below.

(1) x^{-1} (2) x^{-2} (3) x^{-3} (4) x^{-4} (5) x^{-5} (6) x^{-6} (7) x^{-7}.

Has DERIVE given the answers you predicted? If not, can you reconcile the two sets of answers?

(Hint: remember the index laws, $x^{-n} = \frac{1}{x^n}$).

(C)   Repeat the procedure of (B) for:

(1) $x^{1/2}$

NOTE: You will have to type $x^{(1/2)}$ here! Also $x^{1/2}$ and $\sqrt{x}$ mean the same thing.

(2) $x^{1/3}$ (3) $x^{1/4}$ (4) $x^{2/3}$ (5) $x^{3/4}$

(6) $x^{-4/5}$ (7) $x^{-9/7}$ (8) $x^{-15/11}$.

Is there any problem? Can you reconcile what is on the screen with what you predicted?

From what you have done complete this expression.

ie. $\dfrac{d}{dx}(x^n) = ?$

(D)   Now try differentiating the following expressions:

(1) $3x^2$ (2) $5x^6$ (3) $12x^{-9}$ (4) $14x^{1/2}$ (5) $15x^{-2/3}$

Can you predict what the next results will be before you do them with DERIVE?

(6) $6x^2 + 5x - 1$ (7) $6x^3 - 4x^2 + 3x - 2$

(8) $7x^9 + 8x^{-2} - 18x^{1/3} - 15x^{-2/5} + 7$

Do these results fit in with your general conclusion?

Deduce rules for differentiating sums of functions and expressions of the form cx^n where c is a constant.

Summary

Two important features of differentiation have been introduced in this DERIVE Activity. First the notation for the derivative of a function f is written as

$$\frac{d}{dx}(f(x)) \ .$$

For example, if $f(x) = x^2$ we write the derivative of x^2 as

$$\frac{d(x^2)}{dx} \ .$$

Secondly, the rule for differentiating powers of x is

$$\frac{d(x^n)}{dx} = nx^{n-1}$$

Furthermore in the last part of the investigation you saw that the derivative of cx^n is cnx^{n-1} where c is a constant and the derivative of a sum is the sum of the derivatives. Apply these rules in the following exercise.

Exercise 6A

1.   Differentiate the following functions.

(a) x^7 (b) $x^{1/3}$ (c) x^{-3} (d) $x^{-5/2}$

(e) $4x^3$ (f) $6x^{3/2}$ (g) $\dfrac{0.7}{x}$ (h) $x^9 + x^4 + x$

(i) 5   (j) $6x^5 - 2x + 7$ (k) $3x^{-1} + 2x^{-1/2}$

(l) $15x - 3x^4 + 2x^7$ (m) $8x^3 + x^2 - 3x + 2$ (n) $0.7x^9 - 0.3x^{-2}$
(o) $0.1x^{-3} + 1.9x^3$ (p) c, where c is a constant.

2.   Find the values of the derivative of each given function at the given value of x.

(a) x^3 at $x = 2$ (b) x^5 at $x = 1$,   (c) x^2 at $x = 0.5$

(d) x^4 at $x = -1$ (e) $\dfrac{1}{x}$ at $x = 3$ (f) $x^{1/2}$ at $x = 4$.

DERIVE ACTIVITY 6B

The aim of this investigation is to show what differentiation actually means.

Load the Utility file DIF_APPS using **Transfer**.

(A)   **Author** and **Plot** the graph of the function x^2.
       Author and **Simplify** TANGENT(x^2,x,2).
       Plot the expression you obtain.

       The DERIVE screen should look like Figure 6.1.

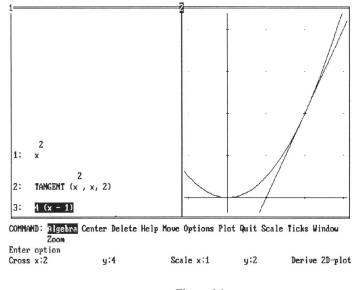

Figure 6.1

The equation of the tangent to the graph is $y = 4x - 4$. The slope of this line is 4. The derivative of x^2 is $2x$ and its value at $x = 2$ is also 4.

(B)   Repeat activity (A) for the function x^2 and other values of x and complete the
       following table.

x	slope of tangent (from DERIVE)	value of derivative (by calculating value of $2x$)
-1		
0		
0.5		
1		
1.5		
2	4	4

What do you conclude from your table of results?

(C)   Repeat activity (B) for the functions x^3, x^4, $x^{1/2}$, x^{-1}. For each function construct
       a table of your results. Note that you will need to use a different selection of
       values of x for $x^{1/2}$ and x^{-1}.

value of x	slope of tangent	value of derivative

Does this confirm your conclusion in (B)?

Summary

From DERIVE Activity 6B we can deduce the following result.

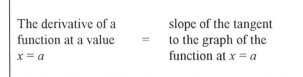

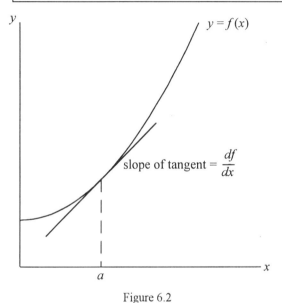

Figure 6.2

Example 6A

Find the equation of the tangent to the graph of the function $4x^3 + 2x - 8$ at the point $(1,-2)$.

Solution

The tangent is a straight line and the general equation of a straight line is $y = mx + c$. The slope of the tangent m at the point $(1,-2)$ equals the value of the derivative of the function at $x = 1$.

$$\frac{d}{dx}(4x^3 + 2x - 8) = 12x^2 + 2$$

and when $x = 1$, the derivative $= m = 14$.

The tangent has equation $y = 14x + c$.
Since the tangent passes through the point $(1, -2)$ we have

$$-2 = 14 + c$$
$$c = -16$$

The equation of the tangent is $y = 14x - 16$.

Exercise 6B

1.　Find the slope of the tangent to the graphs of each of the following functions at the given values of x. Hence find the equations of the tangent at each point.

(a)　x^2 at $x = 3$　　　　(b)　x^3 at $x = -2$　　　　(c)　x^5 at $x = 0.5$

(d)　$x^{1/2}$ at $x = 4$　　　(e)　$x^{5/4}$ at $x = 1$　　　(f)　$x^{0.7}$ at $x = 4$

(g)　x^{-1} at $x = 0.3$　　(h)　$x^{-1.9}$ at $x = 1.2$　　(i)　$3x^4$ at $x = -2$

(j)　$2x^{2.5}$ at $x = 9$　　(k)　$3x^2 - 2x + 1$ at $x = 0$　　(l)　$5x^{-1} - 2x$ at $x = 1$

2.　Find the slope of the tangent to the graphs of each of the following functions at the given values of x. Deduce whether the functions are increasing or decreasing at the given values.

(a)　$4x^3 + 2x^2 - x$　　　　　　$x = 0, x = 1$.

(b)　$x^3 - 3x^2 + 3x - 1$　　　　$x = -1, x = 1, x = 2$.

(c)　$5x^{-1} - 2x$　　　　　　　　$x = 1, x = 2$.

(d)　$x^{1.3} + x^{-0.7}$　　　　　　$x = 0.5, x = 1.5$.

What can you deduce about the shape of the graph between $x = 0.5$ and $x = 1.5$.

6.3   RATES OF CHANGE

In this section we introduce a formal definition of differentiation. The idea of a rate of change as a speed or an acceleration is probably familiar to you. As an example of the ideas involved consider a recent journey between two of the campuses of the University of Plymouth in Plymouth and Exmouth, a distance of 50 miles. The journey out of Plymouth was pretty slow with a high density of traffic and a speed limit. On the A38 between Plymouth and Exeter we travelled at a steady 70mph only slowing down for road works near Newton Abbot. The last part of the journey from the M5 to Exmouth is along country lanes where once again speed is restricted. The journey took 1¼ hours. What can we say about the speed of the journey? Clearly the speed given by the speedometer changed many times (probably continuously) during the trip. It would have been 0mph at traffic lights and up to 70mph on the motorway. Looking at speed more globally, since the 50 mile journey took 1¼ hours, the *average speed* is just 50/1¼ = 40mph. An average speed could be used to give us a rough idea of the time for a journey. Suppose I travel 100 miles along similar roads to the trip above. I might expect an average speed of 40mph and a journey time of 2½ hours. The speedometer would give some idea of the speed at an instant so this is called *the instantaneous speed*. (It is this speed that the police would be interested in if you travel at 70mph through a built up area!!). Since

$$\text{average speed} = \frac{\text{distance travelled}}{\text{time taken}}$$

we define this as the *average rate of change* of distance with time. Over a very short interval of time the average speed will provide a good approximation to the instantaneous speed. The smaller the time interval the better the approximation. In mathematical language we would write

$$\text{instantaneous speed} = \lim_{t \to 0} \frac{s}{t}$$

where s is the distance travelled in time t. A natural question is what does this mean? As t gets smaller so does s and in the limit we have 0/0!

MATHEMATICAL INVESTIGATION 6A

A small object slides along a table so that the distance from its starting point (in metres) is modelled by

$$s = -1.3t^2 + 7.8t$$

where t is the time in seconds.

The object comes to rest after 3 seconds.

(i)   Calculate the total distance travelled and the average speed for the journey.

(ii)   Calculate the average speed during the 1st second ($t = 0$ to 1), during the 3rd second of the journey ($t = 2$ to 3).

(iii) By completing the following table find the instantaneous speed when $t = 1$.

t	s	distance travelled between $t = 1$ and t	average speed between $t = 1$ and t
1.2	7.488	$7.488 - 6.5 = 0.988$	$0.988/0.2 = 4.94$
1.1			
1.01			
1.001			
1.0001			

Estimate the instantaneous speed at $t = 1$.

(iv)   By choosing appropriate times repeat part (iii) to estimate the instantaneous speed at time $t = 2$ seconds.

(v)   Find the value of $\dfrac{ds}{dt}$ at $t = 1$ and $t = 2$.

What can you deduce from your results?

This activity suggests $\lim\limits_{t\to 0}\dfrac{s}{t}$ does converge and that the instantaneous speed is associated with the derivative of the distance function.

Consider a more formal algebraic approach. Let $s(t) = -1.3t^2 + 7.8t$ and suppose we wish to evaluate the speed at $t = t_0$. For a small time h the distance travelled between $t = t_0$ and $t = t_0 + h$ is

$$s(t_0 + h) - s(t_0) = \left[7.8(t_0 + h) - 1.3(t_0 + h)^2\right] - [7.8t_0 - 1.3t_0^2] = 7.8h - 2.6t_0 h - 1.3h^2 \ .$$

The average speed during the time interval h is

$$\text{average speed} = \frac{\text{distance travelled}}{\text{time taken}}$$

$$= \frac{7.8h - 2.6t_0 h - 1.3h^2}{h}$$

$$= 7.8 - 2.6t_0 - 1.3h \ .$$

As the value of h decreases, the distance travelled decreases but the average speed tends to $7.8 - 2.6t_0$.

We deduce that the instantaneous speed at time t_0 is $7.8 - 2.6t_0$.

Notice that the derivative of $s(t)$ is $7.8 - 2.6t$ so that at time $t = t_0$

$$\text{instantaneous speed} = \text{rate of change of distance}$$
$$= \text{derivative of distance function}\ .$$

For an object moving so that its distance function is modelled by $s(t)$ at time t, the instantaneous speed of the object v is given by

$$v = \frac{ds}{dt} \ .$$

The acceleration is related to the speed in a similar way. We have

$$\text{average acceleration} = \frac{\text{change in speed}}{\text{time}}$$

and

$$\text{instantaneous acceleration} = \text{rate of change of speed}$$

$$= \lim\limits_{\text{time}\to 0} \frac{\text{change in speed}}{\text{time}} = \frac{dv}{dt} \ .$$

Exercise 6C

1. A car travels from London to Leeds, a distance of 180 miles in 4½ hours. Calculate the average speed for the journey.
   If most of the journey takes place on the M1, suggest reasons why the average speed is not nearer 70 mph.
   Sketch a possible graph for the distance travelled against time assuming that the driver stops at two motorway service stations.

2. The speed of an object rolling on a sloping table changes from 3.1 ms^{-1} to 4.6 ms^{-1} in 1.3 seconds. Calculate the average acceleration of the object during this time interval.

3. A stone is thrown up in the air so that its height above the thrower is modelled by

   $$h = 40t - 5t^2$$

   where t is time in seconds.

   (a) Find the average speed of the stone during the time intervals

   (a) $t = 0$ to 1   (b) $t = 1$ to 2   (c) $t = 2$ to 2.5.

   (b) Find the instantaneous speed of the stone at times $t = 0, 1, 2, 3$ and 4 seconds. What happens to the stone at $t = 4$?

   (c) Find the instantaneous acceleration of the stone at times $t = 0, 1, 2, 3$ and 4 seconds. What does the negative sign mean?

4 The population of a species of fish in a lake is modelled by

   $$P = 4\sqrt{t} + 3$$

   where t is the time in seconds.

   Calculate the (instantaneous) rate of change of the population at time $t = 2$ and $t = 9$.

5.   The area of an ink blot is given in terms of its radius by

$$A = \pi r^2 .$$

Find the rate of change of the area with respect to r.

6.   A function f is defined by $f(x) = 3x^2 - 4x + 6$.

(a)   By completing the following table estimate the instantaneous rate of change of f when $x = 1.5$.

x	$f(x)$	change in $f(x)$ between x and 1.5	average rate of change between x and 1.5
1.7			
1.6			
1.55			
1.51			
1.501			
1.5001			

(b)   Check your estimate using differentiation.

7.   A function p is defined by $p(r) = 3r - 4/r$.
By completing the following table estimate the instantaneous rate of change of p when $r = 2$.

r	$p(r)$	change in $p(r)$ between r and 2	average rate of change between r and 2
2.2			
2.1			
2.01			
2.001			
2.0001			

Limit Definition of Differentiation

Consider a general function $f(x)$ and its change between two values of x, $x_0 + h$ and x_0. The average rate of change of f is defined by

$$\text{average rate of change} = \frac{f(x_0 + h) - f(x_0)}{h}.$$

On a graph of f this average rate of change is the slope of a chord between points $(x_0, f(x_0))$ and $(x_0 + h, f(x_0 + h))$ (see Figure 6.3).

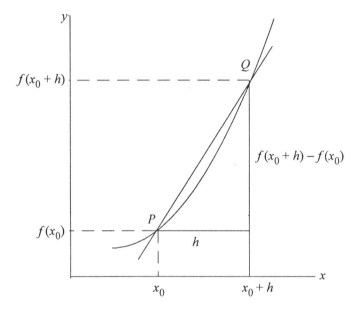

Figure 6.3

As the size of h is reduced the point Q moves round the curve towards point P. The slope of the chord approaches the slope of the tangent to the curve at P (see Figure 6.4).

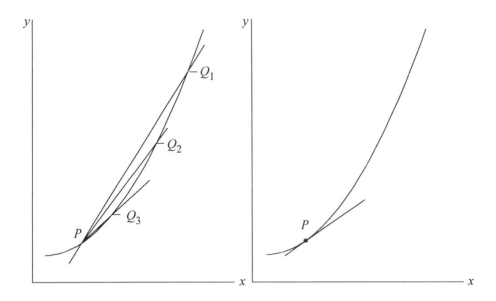

Figure 6.4

We have seen that the slope of the tangent at P is the value of the derivative of the function at P. So now we have a formal link between rates of change and the derivative.

Definition of the derivative of a function $f(x)$

$$\frac{df}{dx} = \frac{d}{dx}(f(x)) = \lim_{h \to 0} \frac{f(x+h) - f(x)}{h}$$

DERIVE ACTIVITY 6C

The aim of this investigation is to demonstrate the link between the limit definition and differentiation.

(A)   **Author** $(x^2-1)/(x-1)$. The brackets are essential! You will see on the screen

$$\frac{x^2-1}{x-1}.$$

Select **Calculus Limit** and choose variable:x. You are prompted to enter the limit point on the real number line and the direction from which it is approached. Enter 1 and Above. Now on screen you should see:

$$\lim_{x \to 1+} \frac{x^2-1}{x-1}.$$

N.B. $x \to 1+$ will mean x tends to 1 from above.

Simplify to obtain the answer 2.

If you put $x = 1$ in the original expression and simplify you will see ?. This means that the value of the expression is undefined; however the limit value tells us that the graph of the function passes through the point whose coordinates are (1,2). Now highlight the original expression and **Plot** it.

In the Algebra window, for the original expression issue the commands **Calculus Limit** 1 Below. On the screen you should see:

$$\lim_{x \to 1-} \frac{x^2-1}{x-1}.$$

N.B. $x \to 1-$ will mean x tends to 1 from below.

Simplify and you will obtain the answer 2 once again. This means that there is no difference in the limiting value whether you approach 1 from above or from below. The graph confirms this.

(B)   Repeat the procedure for each of the limits that follow. You need to evaluate
each limit twice, once from above and once from below to see if there is any
difference.

For clarity do not have more than one graph at a time in the plot window.

The most convenient scales for each graph are given for you. Set these up with
the **Scale** command after moving to the plot window but before plotting the
graph. Clean the windows each time.

	Expression		Suggested x and y Scales	
(i)	$\lim\limits_{x \to 4} \dfrac{x^2 - 16}{x - 4}$		$x{:}3$	$y{:}3$
(ii)	$\lim\limits_{x \to 3} \dfrac{x^3 - 8x - 3}{x - 3}$		$x{:}3$	$y{:}10$
(iii)	$\lim\limits_{x \to 0} \dfrac{\sin x}{x}$		$x{:}10$	$y{:}1$
(iv)	$\lim\limits_{x \to 0} \dfrac{\sin x}{x^2}$		$x{:}10$	$y{:}0.1$
(v)	$\lim\limits_{x \to 0} \dfrac{1 - \cos x}{x}$		$x{:}10$	$y{:}1$
(vi)	$\lim\limits_{x \to 0} \dfrac{\sin x - x \cos x}{x^3}$		$x{:}10$	$y{:}0.2$
(vii)	$\lim\limits_{x \to 0} \dfrac{4^x - 1}{x}$		$x{:}11$	$y{:}1$
(viii)	$\lim\limits_{x \to 0} x^x$		$x{:}2$	$y{:}2$
(ix)	$\lim\limits_{x \to 3} \dfrac{1}{x - 3}$		$x{:}5$	$y{:}5$

Which of them gives different answers on approaching from above and from
below? Does the graph show you why?

(C)   Evaluate the following limits.

(i) $\displaystyle \lim_{n \to \infty} \left[1 + \frac{1}{n} \right]^n$ Limit variable n. (Type inf for ∞).

(ii) $\displaystyle \lim_{n \to \infty} \left[1 + \frac{x}{n} \right]^n$ Limit variable n.

Do you recognise the results?

(D)   Close down the plot window using the **Window Close** commands and **Remove** your expressions.

We have seen that the definition of a derivative involves a limit.

$$\frac{df}{dx} = \lim_{h \to 0} \frac{f(x+h) - f(x)}{h} .$$

Author $((x+h)^2 - x^2)/h$. (Be very careful with the brackets).

Issue the instructions **Calculus Limit**: h 0 **Both**.

On screen you should see:

$$\lim_{h \to 0} \frac{(x+h)^2 - x^2}{h} .$$

Simplify the expression and you should get the result $2x$.

What is the derivative of x^2. The result of differentiating should be the same as the result of the limiting process.

(E)   Repeat the procedure in (D) for the expressions that follow.

Use **Window Split Vertical** at 40 to create a second Algebra window. Switch between them with F1. Use Window 1 for the limits and Window 2 for the derivatives.

Remember that the limit variable is h tending to 0 from above, but the differentiation variable is x order 1.

(i) $\displaystyle \lim_{h \to 0} \frac{(x+h)^3 - x^3}{h}$

$\dfrac{d(x^3)}{dx}$

(ii) $\displaystyle \lim_{h \to 0} \frac{(x+h)^4 - x^4}{h}$

$\dfrac{d(x^4)}{dx}$

(iii) $\displaystyle \lim_{h \to 0} \frac{\left\{4(x+h)^3 - 2(x+h)^2 + 7\right\} - \{4x^3 - 2x^2 + 7\}}{h}$

$\dfrac{d}{dx}(4x^3 - 2x^2 + 7)$

(iv) $\displaystyle \lim_{h \to 0} \frac{\left\{13(x+h)^7 - 5(x+h)^4 + 13(x+h)\right\} - \{13x^7 - 5x^4 + 13x\}}{h}$

$\dfrac{d}{dx}(13x^7 - 5x^4 + 13x)$

(v) $\displaystyle \lim_{h \to 0} \frac{\left\{4(x+h)^{-\frac{1}{2}} + \dfrac{3}{(x+h)}\right\} - \left\{4x^{-\frac{1}{2}} + \dfrac{3}{x}\right\}}{h}$

• $\dfrac{d}{dx}\left(4x^{-\frac{1}{2}} + \dfrac{3}{x}\right)$

Choose other functions and check that the limit process gives the same result as differentiation.

Example 6B

A function $f(x)$ is given by

$$f(x) = 4x^3 + 2x^2 - 3x - 7 .$$

By evaluating $\displaystyle\lim_{h \to 0} \frac{f(x+h) - f(x)}{h}$ confirm the limit definition of the derivative.

Solution

$$\begin{aligned}
f(x+h) &= 4(x+h)^3 + 2(x+h)^2 - 3(x+h) - 7 \\
&= 4x^3 + 12x^2h + 12xh^2 + 4h^3 + 2x^2 + 4xh + 2h^2 - 3x - 3h - 7
\end{aligned}$$

$$f(x+h) - f(x) = 12x^2h + 12xh^2 + 4h^3 + 4xh + 2h^2 - 3h$$

Dividing both sides by h,

$$\frac{f(x+h) - f(h)}{h} = 12x^2 + 12xh + 4h^2 + 4x + 2h - 3 .$$

As h tends to zero this ratio becomes

$$\lim_{h \to 0} \frac{f(x+h) - f(x)}{h} = 12x^2 + 4x - 3 .$$

Using the rules of differentiation of powers of x we have

$$f(x) = 4x^3 + 2x^2 - 3x - 7$$

$$\frac{df}{dx} = 12x^2 + 4x - 3 .$$

Confirming the formal definition of differentiation.

Summary

We now bring together the ideas of differentiation introduced so far.

The **derivative** of a function $f(x)$ with respect to x is written as $\dfrac{df}{dx}$ and has the following meanings:

1.   the derivative is the (instantaneous) **rate of change** with respect to x;

2.   the derivative is the **slope of the tangent** to the graph of $f(x)$;

3.   formally $\dfrac{df}{dx} = \lim_{h \to 0} \dfrac{f(x+h) - f(x)}{h}$;

4.   if $f(x) = x^n$ then $\dfrac{df}{dx} = nx^{n-1}$.

An alternative notation for the derivative of a function $f(x)$ with respect to x is $f'(x)$.

We shall use both notations $\dfrac{df}{dx}$ and $f'(x)$ in subsequent chapters.

Exercise 6D

1.   Confirm the link between $\lim_{h \to 0} \dfrac{f(x+h) - f(x)}{h}$ and the rule for differentiating powers of x for the following functions.

   (a) x^2 (b) x

   (c) $3x^2 - 4x$ (d) $5x^2 + 2x - 8$

   (e) x^3 (f) $2x^3 + 3x - 1$

   (g) $\dfrac{1}{x}$ (h) c, where c is a constant.

2.   In problem 1 the independent variable is x. What would be the formal limit definition for the derivative of a function $g(t)$ where t is the independent variable? Use your formula for the following functions.

   (a) t (b) $3t^2 - 4t + 1$

   (c) $5t^3 - 6t$ (d) 4

6.4   MAXIMUM AND MINIMUM VALUES

Features of Graphs of Functions

Figure 6.5 shows the graphs of two quadratic functions $x^2 - 4$ and $4 - x^2$.

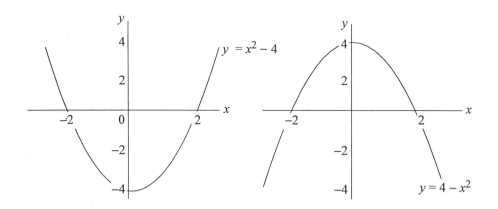

Figure 6.5

If you were asked to describe the graphs describing their essential features you would probably include the following points.

1. They each cut the x axis at two points $x = -2$ and $x = 2$.

2. The graph of $f(x) = x^2 - 4$ has a minimum at the point $(0,-4)$. It continuously decreases up to the point $(0,-4)$ and then increases as x continues to increase.

3. The graph of $f(x) = 4 - x^2$ has a maximum at the point $(0,4)$. It increases until the point $(0,4)$ and then decreases as x continues to increase.

To describe the graph of a function we note the features as the direction of x increases, ie. we scan the graph moving from left to right.

DERIVE ACTIVITY 6D

The aim of this investigation is to explore the important features of a graph of a function. **Load** the **Utility** function DIF_APPS using **Transfer**.

(A)   (i)   **Author** and **Plot** the function

$$f(x) = x^5 - 3.75x^3 - 1.25x^2 + 3.75x + 3.5$$

using scales $x{:}1$ $y{:}5$.

Describe the important features of the graph.
Use the TANGENT function of DERIVE to draw the tangent to the graph at $x = -1.5, -1, -0.5, 0.5, 1.0, 1.5$ and 2.
Make a note of the slope of the tangent at each of these points.

Figure 6.6 shows the DERIVE plot of this function.

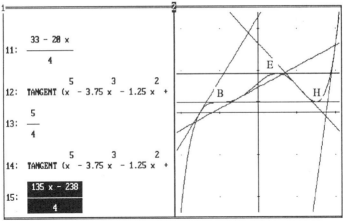

Figure 6.6

At points B, E and H the slopes of the tangents are zero. We call these

points *turning points* or *stationary points*. At these points $\dfrac{df}{dx} = 0$.

(ii)   Use the **Calculus Differentiate** command to find $\dfrac{df}{dx}$ and show that

$\dfrac{df}{dx} = 0$ at $x = -1$, 0.5 and 1.5.

The shape of the curve is quite different at points B, E and H.

As you approach B by moving from A to B to C, the slope of the tangent is positive at A, zero at B and positive at C. The curve crosses the tangent. Point B is an example of *a point of inflexion*.

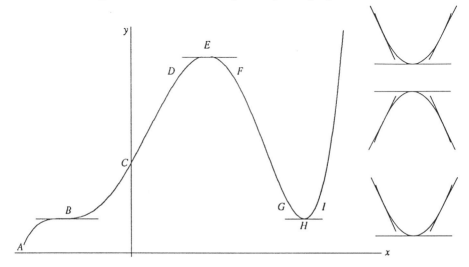

Figure 6.7

(iii)   Discuss whether there is another point of inflexion with the property that the curve crosses the tangent.

At point E the slope changes from a positive value at D to a negative value at F with the rate of change being 0 at E.
Point E is an example of *a local maximum*.

Point H is an example of a *local minimum* at which the slope changes from a negative value at G to a positive value at I with the rate of change being 0 at H.

(B)   Clear the algebra and graphics windows. **Author** and **Plot** the following
       functions.

   (i) $f(x) = x^3 - 4x$

   (ii) $f(x) = 3x^4 + 4x^3 - 24x^2 - 48x - 5$

   From your graphs describe the features of each function. Show that the
   derivative is zero at each of the turning points.

(C)   The function $f(x) = x^2 - 3x + 2$ is the derivative of some unknown function.
       Author and **Plot** this function.
       Describe the properties of the unknown function and propose a graph of this
       function.

Example 6C

Find the turning points of the function $f(x) = x^4 - 4x^2 + 2$ and describe them.

Solution

DERIVE would provide a graphical means of describing the properties of this
function. In this example we will use an algebraic approach and calculus.

The turning points are given by $\dfrac{df}{dx} = 0$.

Differentiating f we have

$$\frac{df}{dx} = 4x^3 - 8x = 4x(x^2 - 2) \ .$$

Solving for $4x(x^2-2) = 0$ for x gives:

$$x = 0 \ \text{ or } \ x = -\sqrt{2} \ \text{ or } \ x = \sqrt{2} \ .$$

There are three turning points. To classify them we look at the slope of the tangent
(ie. the value of the derivative) either side of the turning point.

x	$\dfrac{df}{dx}$	shape of graph	classification
−1.5	−1.5		
−$\sqrt{2}$	0		local minimum
−1	4		
−0.5	3.5		
0	0		local maximum
0.5	−3.5		
1	−4		
$\sqrt{2}$	0		local minimum
1.5	1.5		

With this information we can sketch a graph of the function. This is shown in Figure 6.8.

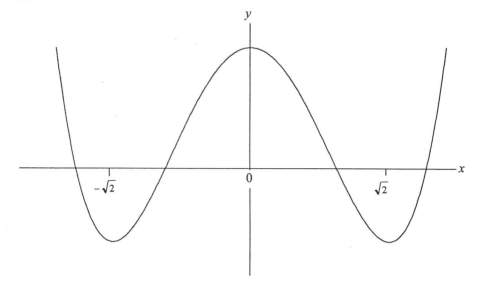

Figure 6.8

Exercise 6E

1.   Find the turning points of the following functions and classify them. Sketch
      the graphs of the functions.

   (a) $3x^2 - 12$ (b) $9 - x^2$

   (c) $2x^2 + 5x - 3$ (d) $2 - x - 3x^2$

   (e) $x^3 + 3$ (f) $x^3 - 5x^2 - x + 5$

   (g) $5 - 12x - 2x^2 + 4x^3 - x^4$ (h) $x + \dfrac{1}{x} - 2$

2.   Use DERIVE to draw a graph for each of the functions in problem 1.
      Use your graphs to investigate whether any of the functions have a point of
      inflexion.

3.   Consider the following graphs of two functions.
      Show where the function is

      (a) increasing,   (b) decreasing,   (c) stationary.

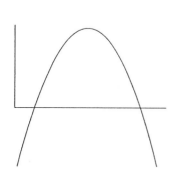
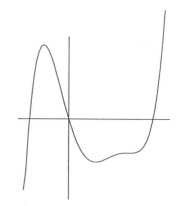

Figure 6.9

4.   You are given the following information about a function $f(x)$

(i) $f(x) = 1$ when $x = 0$;

(ii)   at the points (1,2), (3,–2) and (5,3) the graph is "stationary";

(iii) $f(x) = 0$ when $x = -\frac{1}{2}$, when $x = 2$, when $x = 4$ and when $x = 6$;

(iv)   the gradient is positive when $x = 0$ and when $x = 4$;

(v)   the gradient is negative when $x = 2$ and when $x = 6$.

Use all this information to draw a rough sketch graph of the function.

The Second Derivative

When a function f is differentiated a new function is obtained. This new function is called the *derived function*. For example, if $f(x) = x^3 - 3x^2 + 1$ the derived function is

$$\frac{df}{dx} = 3x^2 - 6x.$$

We can differentiate this function to obtain

$$\frac{d}{dx}(3x^2 - 6x) = 6x - 6 .$$

So we can think of the function $6x - 6$ as the result of having differentiated the initial expression $x^3 - 3x^2 + 1$ twice. This new expression $6x - 6$ is called the *second derivative* of the function f and is written as

$$\frac{d}{dx}\left(\frac{df}{dx}\right) = \frac{d^2 f}{dx^2} = f''(x) .$$

Example 6D

Find the first derivative and second derivative of the function $f(x) = x^4 - 3x^3 + 4x - 7$.

Solution

The first derivative is just the derived function

$$\frac{df}{dx} = 4x^3 - 9x^2 + 4 \ .$$

To find the second derivative we just differentiate again

$$\frac{d^2 f}{dx^2} = \frac{d}{dx}(4x^3 - 9x^2 + 4) = 12x^2 - 18x \ .$$

Exercise 6F

1.   Find the second derivative of each of the following functions.

    (a) $\ x^3 - 3x$ (b) $\ 4x - 2x^3 + 5x^5 - x^6$

    (c) $\ x^2 - x^{\frac{1}{2}} + x^{-\frac{1}{4}}$ (d) $\ x + \dfrac{1}{x}$

    (e) $\ 2\sqrt{x} + x^3$ (f) $\ 3x + 4x^2 - 7x^6$

    (g) $\ 4t^2 - 3t + 7$ (h) $\ 11t - 4$

    (i) $\ t^{\frac{1}{2}} - 2t^{\frac{1}{4}} + 3t^{-\frac{1}{3}}$ (j) $\ x^n$ where n is a constant.

2.   Investigate the second derivatives of functions using DERIVE.

3.   We can continue the process, so that the third derivative of f is the derivative of $f''(x)$ and is written $f'''(x)$. Find the second, third and fourth derivatives of the following functions.

    (a) $\ x^5 - 2x^4 + 3x$ (b) $\ 4x - 2x^3 + 5x^5 - x^6$

    (c) $\ \dfrac{1}{x}$ (d) $\ \sqrt{t} - 2t^3$

DERIVE ACTIVITY 6E

The aim of this investigation is to relate values of the second derivative of function f to the graph of the function f.

Open three graphics windows, so that window 2 shows the graph of a function $f(x)$, window 3 shows the graph of the derivative of f and window 4 shows the graph of the second derivative of f. Use the F1 key to move between windows.

(A)   In window 1 **Author** the function

$$f(x) = x^3 - 3x^2 - x + 3 \ .$$

Plot this in window 2.

Differentiate $f(x)$ and plot the graph of $\dfrac{df}{dx}$ in window 3.

Differentiate $\dfrac{df}{dx}$ and plot the graph of $\dfrac{d^2 f}{dx^2}$ in window 4.

Your DERIVE screen should look like Figure 6.10.

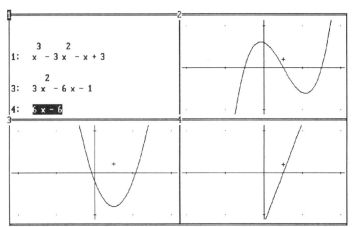

Figure 6.10

Check the following statements from your graph.

1.   The zeros of the derivative coincide with the local maximum and local minimum of f.

2.   At the local maximum the second derivative is negative.

3.   At the local minimum the second derivative is positive.

4.   At the point of inflexion the second derivative is zero.

(B)   Investigate these statements with the following functions.

(i) $f(x) = x^3 + x^2 - 6x$ (iii) $f(x) = x^4$

(ii) $f(x) = x^4 - 5x^2 + 4$ (iv) $f(x) = -2x^3 + x^2 + 7x - 6$

Discuss how values of the second derivative can be used to explore the properties of a function.

Summary

The second derivative represents the rate of change of the tangent to the graph of a function, and its value at the turning points can be used to classify them.
At a local maximum:

$$\frac{df}{dx} = 0 \text{ and } \frac{d^2 f}{dx^2} < 0$$

(the slope of the tangent is decreasing from positive to negative values and so the rate of change of tangent is negative).
At a local minimum

$$\frac{df}{dx} = 0 \text{ and } \frac{d^2 f}{dx^2} > 0$$

(the slope of the tangent is increasing from negative to positive values and so the rate of change of the tangent is positive).

At a point of inflexion $\dfrac{d^2 f}{dx^2} = 0$. If the first derivative is also zero at a point of inflexion the graph may look like Figure 6.11.

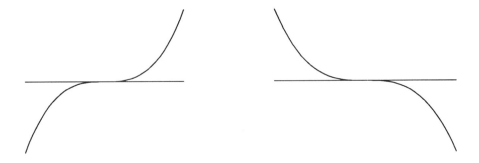

Figure 6.11

You should use these results with care when classifying the turning points of a function. For example, consider the functions $f(x) = x^3$ and $f(x) = x^4$.

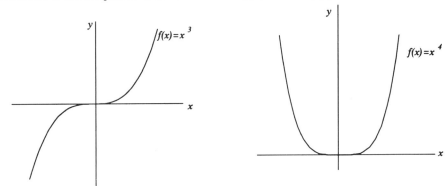

Figure 6.12

For $f(x) = x^3$, $\dfrac{df}{dx} = 0$ and $\dfrac{d^2 f}{dx^2} = 0$ at $x = 0$.

For $f(x) = x^4$, $\dfrac{df}{dx} = 0$ and $\dfrac{d^2 f}{dx^2} = 0$ at $x = 0$.

However notice the different shapes of the two graphs; $f(x) = x^3$ has a point of inflexion at $x = 0$ whereas $f(x) = x^4$ has a local minimum at $x = 0$.

   If the first and second derivatives of a function are both zero at a value of x, it is necessary to look at the slope of the tangent either side of the turning point in order to classify it.

Exercise 6G

Find the coordinates of the turning points of the following functions. Where appropriate use the second derivative to classify them.

(a) $4x^2 - 11$ (b) $5x^3 - 15x^2 + 15x + 2$

(c) $x^4 - 6x^3 + 11x^2 - 6x$ (d) $x^3 - 6x^2 + 11x - 6$

(e) $x + \dfrac{1}{x}$ (f) $x^2 - \dfrac{1}{x^2}$

Use DERIVE to sketch the graphs of the functions to check your answers.

Applications

In this section we have explored the application of the derivative to investigate the properties of the graphs of functions. The ideas are also useful in solving problems involving the maximum and minimum values of physical quantities.

Example 6E

A carton is being designed to hold ½ litre of milk. The carton must have a square base to fit neatly onto supermarket shelves.

Find the dimensions of the carton so that the material to be used is minimised.

Solution

Step 1   Formulate the mathematical problem

The first step in problems of this type is to set up the function to be minimised (or maximised in other problems). Assume that the carton is to be a cuboid with base size x and height h (in cm).

Volume of box $= x^2h \text{ cm}^3$

The volume of ½ litre of milk is 500 cm³.

Hence

$$x^2h = 500 \qquad (1)$$

The area of material to be used is

$$A = 2x^2 + 4xh \qquad (2)$$

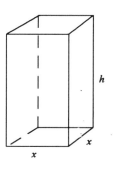

Figure 6.13

(Here we have ignored the material needed to form the flaps).

From equation (1)

$$h = \frac{500}{x^2} \; .$$

Substituting into equation (2) we have

$$A = 2x^2 + \frac{2000}{x} \; .$$

Step 2 Solve the mathematical problem

The amount of material is a minimum when $\dfrac{dA}{dx} = 0$. So differentiating $A(x)$ with respect to x

$$\frac{dA}{dx} = 4x - \frac{2000}{x^2} = 0 \; .$$

Solving for x,

$$4x^3 = 2000$$
$$x^3 = 500$$
$$x = \sqrt[3]{500} = 7.94 \, \text{cm}$$

Substituting into equation (1),

$$h = \frac{500}{x^2} = 7.94\,\text{cm}.$$

The carton with minimum surface area is a cube of size 7.94 cm.

Note in solving realistic problems it is often necessary to make simplifying assumptions so that you can proceed. It is important to list these assumptions carefully.

Exercise 6H

1. A beam has bending moment M (kNm) given by $M = 6x^2 - 12x$ where x is the distance from one end in metres. Find the position of the minimum value of M.

2. A cylindrical fuel storage tank must hold 15 000 litres.
    Find its dimensions if its surface area is to be minimised. What is the area of metal used in constructing the tank.

3. In an aqueous solution the product xy of the concentrations x and y of OH⁻ and H^+ ions respectively is a constant at constant temperature. Deduce under what conditions $x + y$ will be a minimum. Find the concentration of H^+ under minimum conditions at 293°K when $xy = 10^{-4}\,\text{Kmol}^2\text{dm}^{-6}$.

4. For a belt drive the formula relating the power transmitted P to the speed of the belt v is given by

    $$P(v) = Tv - av^3$$

    where T is the tension in the belt and a is a constant.

    (a) Find the speed such that the belt delivers maximum power.
    (b) Sketch the graph of the function $P(v)$.

5. Find the maximum area that can be fenced off from a rectangular field by 80m of fencing using one of the existing walls of the field.

6. A cylindrical tin can without a lid is made of sheet metal. If A is the surface
    area of the sheet used and V is the volume of the can, show that

    $$V = \tfrac{1}{2}(Ar - \pi r^3)$$

    where r is the radius. (Assume that there is no wastage).

    If A is given, show that the volume is a maximum when the diameter of the can
    is twice the height of the can.

7. A piece of wire forms the circumference of a circle of radius 0.16 m. The wire
    is cut and bent to form two new circles. Find the radius of each circle so that
    the sum of the areas of the two circles is a minimum.

8. A cylinder is such that the sum of its diameter and height is 20 cm. Write the
    volume (V cm^3) in terms of the radius of its base (r cm). What is the greatest
    possible value for the volume of the cylinder?

6.5   DIFFERENTIATION OF SPECIAL FUNCTIONS

You know how to differentiate powers of x such as x^2, x^3, x^{-1}, $x^{-\frac{1}{2}}$ etc. But what
about the special functions e^x, $\ln(x)$, $\sin(x)$ and $\cos(x)$? The formal definition as a
limit would be used to find the general results.

The derivative of the exponential function e^x can be deduced from its
introduction in Chapter 2. In DERIVE Activity 2B you saw that the slope of the
tangent to e^x at each x value is equal to the function value of e^x at the x value. In
symbols we have

$$\frac{d}{dx}e^x = e^x .$$

It is this property of e^x that makes it a very important function in mathematics.
It is the only function whose derived function is exactly the same form as the function
itself.

DERIVE ACTIVITY 6F

The aim of this activity is to investigate the derivatives of the functions e^{ax}, $\ln(ax)$, $\sin(ax)$, $\cos(ax)$ where a is a constant.

(A)   Use the **Limit** command in DERIVE to find the following limits. Compare your answers with the **Differentiation** command

$$\lim_{h \to 0} \frac{e^{(x+h)} - e^x}{h} \ .$$

(Remember to input e^x use Alt e so that you see ê on the screen).

$$\lim_{h \to 0} \frac{\ln(x+h) - \ln(x)}{h} \ .$$

$$\lim_{h \to 0} \frac{\sin(x+h) - \sin(x)}{h}$$

$$\lim_{h \to 0} \frac{\cos(x+h) - \cos(x)}{h}$$

(B)   Use DERIVE to investigate the derivative of e^{ax}, $\ln(ax)$, $\sin(ax)$ and $\cos(ax)$ for values of $a = 2, -3, 0.7, -1.3$ and 4.

From the results of your investigation complete the following table.

Function	Derived Function
e^{ax}	
$\ln(ax)$	
$\sin(ax)$	
$\cos(ax)$	

(C)   Predict the derivatives of the following functions and check them with DERIVE.

$$e^{3x} + \sin 2x - \cos 4x$$
$$e^{-2x} - \cos 3x + \ln(5x)$$
$$7e^{0.4x} + 0.9 \sin(4x)$$
$$-0.8e^{-1.6x} - 3.8 \cos(2.7x)$$

Exercise 6I

1. Find the first and second derivatives of each of the following.

   (a) e^{4x} (b) e^{-7x} (c) $4e^{0.5x}$

   (d) $2e^{-1.3x}$ (e) $\ln(5x)$ (f) $3\ln(2x)$

   (g) $\sin(\pi x)$ (h) $\sin(2x)$ (i) $4.2\sin(3.1x)$

   (j) $\cos(4x)$ (k) $\cos(0.2x)$ (l) $1.5\cos(2\pi x)$

   (m) $0.3e^{0.1x} - 0.7\sin(0.5x)$ (n) $4\cos3x - 3\sin4x$

   (o) $e^{-0.1x} + e^{0.1x}$ (p) $\ln(2.6x) - 6\ln(0.7x)$

2. Find the equation of the tangent to the graphs of each of the following functions at the given values of x.

   (a) $2\cos3x$ at $x = 0.5$ (b) $\ln(x)$ at $x = 1$

   (c) e^x at $x = 2$ (d) $e^{-0.1x} - 0.6\cos3x$ at $x = 0$

3. Show that e^{ax} and $\ln(ax)$ have no turning points or points of inflexion.

4. The temperature at a point on a heated rod varies according to the rule

   $$T = 20 + 100e^{-5t}.$$

   Find the rate of change of temperature at (a) $t = 0$, and (b) $t = 1$.

5. The population of a yeast culture is modelled by

   $$P(t) = 4.3e^{-2.1t}.$$

   Find the rate of change of the population.

6. A particle moves so that its displacement as a function of time t is modelled by

   $$s = 0.3\sin(0.7t).$$

   (a) Find the speed of the particle when $t = 0$ and $t = 1$ seconds.

   (b) Find the acceleration of the particle at these times.

6.6   RULES OF DIFFERENTIATION

Many functions are made up from the basic functions x^n, e^{ax}, $\ln(x)$, $\sin(ax)$ and $\cos(ax)$ by the rules of addition, multiplication, division and composition. For example, the motion of a damped oscillating system can be modelled by

$$x = e^{-2t} \sin(0.6t + 0.7)$$

for appropriate system parameters. This function is the product of e^{-2t} and the composite function $\sin(0.6t + 0.7)$.

DERIVE ACTIVITY 6G

The aim of this activity is to investigate the derivatives of products and composite functions.

(A)   Use DERIVE to investigate the derivative of the following products of functions.

(i) $x^2\sin(x)$ (ii) $e^{2x}\cos(3x)$ (iii) $x^3 e^{4x}$

From the results of your investigation suggest a rule for differentiating the product of two functions $f(x) = g(x).h(x)$. Use your rule to predict the derivatives of the following functions before you do them with DERIVE.

(vi) $x^2\ln(x)$ (vii) $e^x\ln(x)$ (viii) $4\sqrt{x}\sin(5x)$

(B)   Many functions are defined as composite functions of the form $g[h(x)]$, for example, e^{x^2}.

Use DERIVE to investigate the derivative of the following functions.

(i) e^{x^2} (ii) $\sin(4x^3)$ (iii) $(3 - x^2)^8$ (iv) $\cos(\sqrt{x})$

From the results of your investigation suggest a rule for differentiating the composite function $f(x) = g[h(x)]$. Use your rule to predict the derivatives of the following functions before you do them with DERIVE.

(v) $\sin(x^2)$ (vi) $e^{(4x^2 - 1)}$ (vii) $(1+x)^8$

(viii) $(a+bx)^7$ (ix) $\ln(a+bx)$ where a,b are constants   (x) $\sqrt{1+3x^3}$

To differentiate any function we need a set of rules for differentiating combinations of functions. The set of rules are quoted as follows:

Sum of two functions

If $f(x) = g(x) + h(x)$

then $\dfrac{df}{dx} = \dfrac{dg}{dx} + \dfrac{dh}{dx}$.

We have already used this rule often in this chapter.

Product of two functions

If $f(x) = u(x).v(x)$

then $\dfrac{df}{dx} = \left(\dfrac{du}{dx}\right).v + u.\left(\dfrac{dv}{dx}\right)$

This is known as the *Product Rule*

Example 6F

Differentiate (a) $x^2 e^{3x}$ and (b) $e^{3x} \sin(4x)$.

Solution

The derivative of x^2 is $2x$ and the derivative of e^{3x} is $3e^{3x}$. Applying the Product Rule

$$\frac{d}{dx} x^2 e^{3x} = 2x e^{3x} + 3x^2 e^{3x} .$$

The derivative of $\sin 4x$ is $4\cos 4x$. Apply the Product Rule

$$\frac{d}{dx} e^{3x} \sin(4x) = 3e^{3x} \sin 4x + 4e^{3x} \cos 4x .$$

Quotient of two functions

If $f(x) = \dfrac{u(x)}{v(x)}$

then $\dfrac{df}{dx} = \dfrac{v \cdot \dfrac{du}{dx} - u \cdot \dfrac{dv}{dx}}{v^2}$.

This is known as the *Quotient Rule.*

Example 6G

Differentiate $\dfrac{x^2}{\sin 3x}$.

Solution

The derivative of x^2 is $2x$ and the derivative of $\sin 3x$ is $3\cos 3x$. Applying the quotient rule

$$\frac{d}{dx}\left(\frac{x^2}{\sin 3x}\right) = \frac{2x \cdot \sin 3x - 3x^2 \cos 3x}{(\sin 3x)^2}.$$

Composition of two functions

For $f(x) = g[h(x)]$

let $u = h(x)$

so that $f(x) = g(u)$

then $\dfrac{df}{dx} = \dfrac{dg}{du} \cdot \dfrac{du}{dx}$.

This is known as the *Chain Rule.*

Example 6H

Differentiate $f(x) = \sin(x^2)$.

Solution

Suppose we let $u = x^2$, then the function f can be written as two simple functions

$$f(x) = \sin(u) \text{ and } u = x^2 .$$

Now $\dfrac{df}{du} = \cos(u)$ and $\dfrac{du}{dx} = 2x$.

Applying the chain rule

$$\frac{df}{dx} = \cos(u).2x = 2x\cos(x^2) .$$

Example 6I

The rate of increase of the radius of a sphere is 0.6 mm per second. Find the rate of increase of the volume of the sphere when the radius is 20 cm.

Solution

If V is the volume of the sphere when its radius is r then

$$V = \frac{4}{3}\pi r^3 .$$

We need to find the rate of change of volume $\dfrac{dV}{dt}$ given that $\dfrac{dr}{dt} = 0.6\,\text{mms}^{-1}$.

Apply the chain rule to the equation for V,

$$\frac{dV}{dt} = \frac{dV}{dr}\cdot\frac{dr}{dt} = \frac{4}{3}\pi(3r^2).\frac{dr}{dt} = 4\pi r^2\frac{dr}{dt}.$$

Substituting for $r = 20$ and $\dfrac{dr}{dt} = 0.6$ we have

$$\frac{dV}{dt} = 4\pi(20^2)0.6 = 3016\,\text{mm}^3\text{s}^{-1}.$$

Exercise 6J

1.   Apply the rules of differentiation to find the derivatives of the following.

(a) $\sqrt{x}\,e^{2x}$ (b) x^2e^{3x} (c) x^5e^{-2x}

(d) $x\sin(2x)$ (e) $\sqrt{x}\cos(\pi x)$ (f) $5x^3e^{3x}$

(g) $\tan(x)$ (h) $x\ln(x)$ (i) $\dfrac{\sin(2x)}{x^2}$

(j) $\dfrac{\sqrt{x}}{e^{3x}}$ (k) $\dfrac{\cos(3x)}{\sin(2x)}$ (l) $\dfrac{e^{2x}+e^{-2x}}{x^2}$

(m) $(3x-1)^5$ (n) $\sqrt{4x+1}$ (o) $\cos(\pi x-3)$

(p) e^{x^2} (q) $\ln(x^2+1)$ (r) $\ln(3\cos(2x))$

(s) $\sin^2x + \cos^2x$ (t) $x^2e^{-2x}\sin(3x)$ (u) $3\cos(1-4x)$

(v) $\left(6x-\dfrac{1}{x}\right)^3$ (w) $e^{5x}\sin(0.7x)$ (x) $\ln(x^2)$

(y) $\sec x = \dfrac{1}{\cos x}$ (z) $\dfrac{1}{\sqrt{4x+1}}$

2.   Find the turning points, if any, of the following functions. Use this information to sketch the curves of the functions.

(a) $t^2 + \dfrac{1}{t}$ (b) $\dfrac{(t+2)}{(t-6)}$

(c) $\dfrac{x^2}{(1-x^2)}$ (d) $\dfrac{(x-3)}{(x+1)}$

(e) $e^{-3t}\sin(2t)$ (f) $\dfrac{e^x}{2+3e^x}$

(g) $\dfrac{xe^x}{(x+1)}$ (h) $x + \sin(x)$

3.   A sector is cut from a circular sheet of metal of radius r and bent round to form a cone.

(a)   Show that if φ is the angle of the sector that is removed then

$$\varphi = 2\pi - 2\pi\sin\theta$$

where θ is the semi-vertical angle of the cone.

(b)   Show that the volume of the cone is

$$V = \frac{1}{3}\pi r^3 \sin^2\theta\cos\theta \ .$$

(c)   Find the angle of the sector removed in order that the volume of the cone is a maximum.

4.   The radius r cm of a circular ink blot on a piece of blotting paper t seconds after it was first viewed is given by

$$r(t) = 12 - \frac{9}{t} .$$

(a)   Calculate the radius of the blot after 3 seconds.

(b)   Find the time when the blot has radius 3cm.

(c)   Find the rate at which the radius is changing when the radius is 3cm. Is the radius then increasing or decreasing?

(d)   What is the largest value of r?

5.   Find the stationary points of the following functions, and determine their nature:

(a) $\dfrac{x^2 - 2x + 4}{x^2 + 2x + 4}$

(b) $\sin(x) - \cos(x)$

(c) $x + \sin(x)$

(d) $e^x \cos(x)$

6.   The power delivered into the load X of a class A amplifier of output resistance R is given by

$$P(X) = \frac{V^2 X}{(X + R)^2}$$

where V is the output voltage.

(i)   Find the value of X such that P is a maximum.

(ii)   Sketch a graph of $P(X)$ against X.

7.   Frequency stability in the cathode-coupled oscillator can be studied with the aid
      of the correction factor

$$f(\alpha) = 1 - \frac{L}{16C}(\alpha - 1/R)^2$$

where L, R and C are constants.

(i)   Show that the maximum value of f is obtained when $\alpha = 1/R$.

(ii)   Sketch a graph of the function $f(\alpha)$.

8.   By considering the derivative of the function

$$f(x) = \sin(x)\tan(x) - 2\ln[\sec(x)]$$

(a)   show that f steadily increases as x increases from 0 to $\pi/2$,

(b)   show that the graph has no points of inflexion between these limits.

9.   The mass of gas which will flow through an orifice from pressure p_1 to p_0 is
      proportional to

$$x^k \sqrt{(1 - x^{1-k})}$$

where k (< 1) is a constant.

Show that the maximum value of this expression occurs when

$$x = \left(\frac{2k}{1+k}\right)^{\frac{1}{(1-k)}}.$$

7

Integration

7.1 AREAS AS SUMS

How do you find the area of an irregular shaped region?
How do you find the distance travelled by a space rocket when you know its speed at given time integrals?

These are typical problems which involve summations. Consider the following examples.

Example 7A

Find the area between the function $f(x) = e^x$, and the lines $x = 0$, $x = 1$ and $y = 0$.

Solution

The shaded area in Figure 7.1 shows the area to be found.

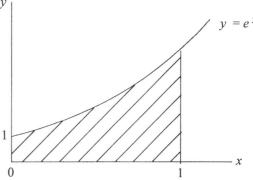

Figure 7.1

The ancient Greeks solved problems
of this type and invented a clever
method of proceeding. They divided
the region into thin strips so that each
strip was rectangular in shape. This
is shown in Figure 7.2.

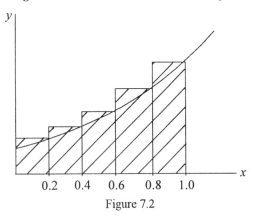

The area of each rectangle is easy to
calculate and the sum of the areas is
then an approximation to the required
area. Clearly the thinner the strips,
the more accurate is the
approximation.

Figure 7.2

Suppose that we take five vertical strips each of equal width 0.2. Then the heights of
the five rectangles are $e^{0.1}$, $e^{0.3}$, $e^{0.5}$, $e^{0.7}$ and $e^{0.9}$. The total area of the five rectangles is
then

$$\text{Area} = (e^{0.1} \times 0.2) + (e^{0.3} \times 0.2) + (e^{0.5} \times 0.2) + (e^{0.7} \times 0.2) + (e^{0.9} \times 0.2)$$
$$= 1.71542 .$$

The area between $y = e^x$, $y = 0$, $x = 0$ and $x = 1$ is approximately 1.715 (to 4 sf). At
this stage we do not know how good is the approximation. By taking double the
number of rectangles so that they each have width 0.1 we get the approximation 1.718
(to 4 sf). (As we shall show later the actual area is 1.71828 to six significant figures).

Example 7B

A rocket is launched into space. The speed of the rocket at two second time intervals
is shown in Table 7.1.

Table 7.1

time (s)	2	4	6	8	10
speed (ms^{-1})	4	16	36	64	100

Estimate the distance travelled during the first ten seconds of the flight.

Solution

In this example we do not know a formula for the speed. Let us assume that during each two second interval the speed of the rocket is constant. The distance travelled during each two second interval is just speed multiplied by time. So we can estimate the distance travelled in ten seconds by summation.

$$\text{Distance travelled} \cong 4 \times 2 + 16 \times 2 + 36 \times 2 + 64 \times 2 + 100 \times 2$$

$$= 440\,\text{m}.$$

This is clearly a simplification based on the assumption of constant speeds. We could improve the approximation by assuming a linear speed in each interval. These two cases are shown in Figure 7.3.

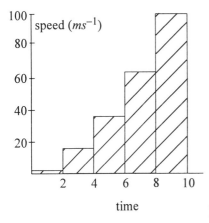 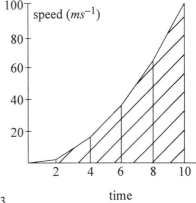

Figure 7.3

Again the summation is an approximation to the area of the region.

More generally consider the problem of finding the area between the graph of $y = f(x)$, $x = a$, $x = b$ and $y = 0$ (see Figure 7.4).

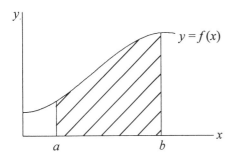 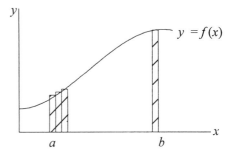

Figure 7.4   Figure 7.5

Suppose that we divide the region into n vertical strips each of the same width $(b-a)/n$ (see figure 7.5). If we assume that the value of the function at the mid-point of each strip represents the height of the rectangle then the expression

$$S_i = f(x_i)\frac{(b-a)}{n}$$

is the area of the ith rectangle with mid-point x_i. The sum of the areas of the n rectangles

$$S = \sum_{i=1}^{n} f(x_i)\frac{(b-a)}{n}$$

is an approximation to the area of the region required.

If we increase the number of rectangular strips, and hence make them thinner, the value of S becomes closer to the value of the required area. In the limit, as $n \to \infty$, the summation should equal the area. Formally we write

$$\text{Area} = \lim_{n \to \infty} \sum_{i=1}^{n} f(x_i)\frac{(b-a)}{n} .$$

The limiting process is a complicated one but for "well behaved continuous functions" it does converge. Finding the limit of a sum is called *integration* and in the above case the *integral of the function f* is an area. We use a special symbol $\int$ to denote it. We write

$$\int_a^b f(x)dx = \lim_{n \to \infty} \sum_{i=1}^{n} f(x_i)\frac{(b-a)}{n} .$$

For example, the area in Example 7A is written formally as $\int_0^1 e^x\, dx$ and the distance travelled by the rocket in Example 7B can be written as $\int_0^{10} v dt$ where $v(t)$ is the velocity function.

Exercise 7A

1.   Estimate the area of the following regions, choosing ten sub-intervals in each
        case.

        (a)   Between $y = x^2$, $y = 0$, $x = 0$ and $x = 3$.

        (b)   Between $y = x^3$, $y = 0$, $x = 1$ and $x = 2$.

        (c)   Between $y = \sin x$, $y = 0$, $x = 0$ and $x = \pi/3$.

        (d)   Between $y = e^{-2x}$, $y = 0$, $x = -1$ and $x = 3$.

2.   Use the $\int$ notation to write each of the areas in problem 1 as an integral.

3.   A particle moves so that its velocity (in ms^{-1}) at time t is given by

$$v = t(8-t^3)/4 \qquad 0 \le t \le 2.$$

        (a)   Sketch a graph of v against t.

        (b)   Estimate the distance travelled during the first two seconds of the motion
                of the particle.

4.   On a car journey the speed of a car is recorded every 5 minutes and is shown in
        the following table. The car starts from rest at $t = 0$.

Table 7.2

time (in min)	5	10	15	20	25	30	35	40
speed (in mph)	20	30	30	15	20	25	15	0

        (a)   Estimate the distance travelled during the journey.

        (b)   How could we obtain a better estimate of the distance travelled?

7.2   EVALUATING INTEGRALS

For two simple functions $f(x) = c$ and $f(x) = mx$ where m and c are positive constants, we can evaluate the integral of f by finding the area summations exactly. For other functions we need a set of standard rules.

Example 7C

Find the integrals of the functions $f(x) = c$ and $f(x) = mx$ between $x = a$ and $x = b$.

Solution

The values of the integrals are shown by the areas in Figure 7.6.

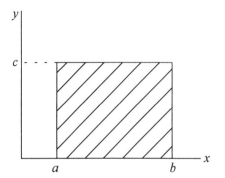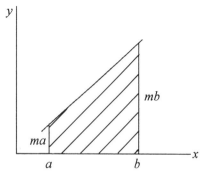

Figure 7.6

(a) $f(x) = c$.

The region is a rectangle whose area is $(b-a)c$; so that

$$\int_a^b c\,dx = (b-a)c \ .$$

(b) $f(x) = mx$.

The region is a trapezium whose area is $\frac{1}{2}(b-a)[ma+mb]$;

$$\int_a^b mx\,dx = (b-a)[\tfrac{1}{2}m(b+a)] = \tfrac{1}{2}m(b^2 - a^2) \ .$$

DERIVE ACTIVITY 7A

The aim of this investigation is to use DERIVE to evaluate the sum and to show how the integral of a function leads to a new function.

Load DERIVE so that you have an algebra window.

Begin by setting up the approximation of Area as a sum by defining the AREA function in the following way.

Declare Function. At the name prompt type f, at the value prompt just press enter and at the variable prompt type x.

$$\textbf{Author } F\left(a+(i-1)\frac{(b-a)}{n}+\frac{(b-a)}{2n}\right).$$

This defines the value of the function at the mid-point of a rectangle x_i since $(b-a)/n$ is the width of a small sub-division.

Use the **Calculus Sum** to form the summation of f with variable i between 1 and n.

Now **Declare Function** AREA with value

$$\frac{(b-a)}{n} * \#3$$

assuming that line #3 has your summation expression.

Your DERIVE screen should look like Figure 7.7.

1:   $F(x) :=$

2:   $F\left[a + \dfrac{(i-1)(b-a)}{n} + \dfrac{b-a}{2n}\right]$

3:   $\displaystyle\sum_{i=1}^{n} F\left[a + \dfrac{(i-1)(b-a)}{n} + \dfrac{b-a}{2n}\right]$

4:   $AREA(a, b, n) := \dfrac{b-a}{n} \displaystyle\sum_{i=1}^{n} F\left[a + \dfrac{(i-1)(b-a)}{n} + \dfrac{b-a}{2n}\right]$

```
COMMAND: Author Build Calculus Declare Expand Factor Help Jump soLve Manage
         Options Plot Quit Remove Simplify Transfer moVe Window approX
Enter option
AREA():=                              Free:96%              Derive Algebra
```

Figure 7.7

(A)   Check the result that the area under the function e^x between 0 and 1 using 5 intervals is approximately 1.71542 by carrying out the following steps.

Declare $F(x)$ to be e^x by entering e^x in response to the **DECLARE FUNCTION** value request (remember alt e for the exponential function).

Author AREA(0,1,5) then **approX**.
The expression is evaluated as 1.71542.
Repeat the approximation by increasing n.
What is the size of n for AREA(0,1,n) = 1.71828?

(B)   Use the expression AREA to estimate the areas of the following regions.

(a)   Between $y = x^2$, $y = 0$, $x = 0$ and $x = 3$.

(b)   Between $y = x^3$, $y = 0$, $x = 1$ and $x = 2$.

(c)   Between $y = \sin x$, $y = 0$, $x = 0$ and $x = \pi/3$.

(d)   Between $y = e^{-2x}$, $y = 0$, $x = -1$ and $x = 3$.

In each case, increase the value of n so that two approximations are equal to four significant figures.

(C)   In this activity you will investigate the relationship between the function f and
the area function.

Clear the screen. Repeat the instructions on page 207 so that you have the
screen in figure 7.7.
Open a graphics window beside your algebra window.
Declare the function $F(x): = x^2$.
Author AREA$(0,x,10)$.

This is the area function evaluated over the interval $[0,x]$.

Plot AREA$(0,x,10)$ and **Plot** x^2.
Return to the **Algebra** window and **Simplify** AREA$(0,x,100)$.
You should obtain the DERIVE screen shown in Figure 7.8.

Use **Calculus Differentiate** to differentiate the function AREA$(0,x,100)$.

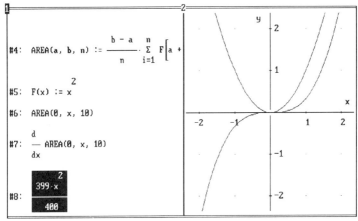

Figure 7.8

The AREA function is clearly a new function of x.

Repeat the activity with the functions x^3 and x^4.

What is the relationship between the AREA function based on x^2 and the
function $f(x) = x^2$?

(D)   Check your conjecture of the relationship between AREA and a function f using
       the following functions

$$4 - x \quad \text{and} \quad 1 - \frac{x^2}{2} + x^4.$$

Summary

The DERIVE Activity 7A has shown that the area function (ie. the integral of a
function) is related to a function f in the following way

$$\frac{d}{dx} \text{AREA}(0, x, 100) = f(x) .$$

Integration is closely related to differentiation and we often think of them as opposite
processes.

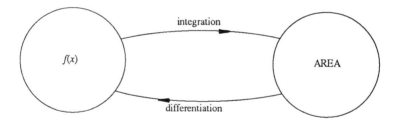

Figure 7.9

DERIVE ACTIVITY 7B

In this investigation you will confirm the link between integration and differentiation and investigate the rules for integrating standard functions.

(A)   **Author** x and press C for **Calculus**.

   Then press I for **Integrate**.
   You will see Integrate expression: #1.
   Press Enter.

   Then you will see CALCULUS INTEGRATE variable:x.
   Again press Enter.

   Now you will be prompted to enter a lower and an upper limit.
   Ignore these, for the present, by pressing Enter.

   On the screen at line #2 you should see the symbols

$$\int x dx$$

   which mean, "Integrate x with respect to the variable x".

   Simplify the expression.

   You should see the result $0.5x^2$ $\left(\text{or } \dfrac{x^2}{2} \right)$.

   Repeat the process for the following functions keeping a record of your results.

$$x^2, x^3, x^4, x^5, x^6, x^7, x^{-1}, x^{-2}, x^{-3}, x^{-4}, x^{-5}, x^{-6}, x^{-7}.$$

   Can you see a pattern emerging?
   You may need to use the index laws to interpret what you see on the screen.

(B)   Try to predict the results of integrating the following functions, then check them using DERIVE.

$$x^{\frac{1}{2}}, x^{\frac{1}{3}}, x^{\frac{1}{4}}, x^{\frac{2}{3}}, x^{\frac{3}{4}}, x^{-\frac{4}{5}}, x^{-\frac{9}{7}}, x^{-\frac{15}{11}}$$

   (Remember you need to use brackets eg. type $x \wedge (1/2)$).

What can you conclude is the general result of integrating x^n with respect to the variable x?

ie. $\int x^n dx = ?$

Is this result true for all values of n? If not, which values do not fit?

(C)   Now investigate the effect of integrating

$$3x^2, 5x^6, 12x^{-9}, 14x^{\frac{1}{2}}, 15x^{-\frac{2}{3}}.$$

Clear the screen before carrying on.

(D)   Differentiate $x^3 + x^2 + x + 1$.
Now integrate the result.
What has happened?

Differentiate $x^3 + x^2 + x + 5$ and integrate the result.
Has the same thing happened again?

Do the same for $x^3 + x^2 + x - 10$.

Repeat the process for

$$6x^2 + 5x - 1, \ 6x^3 - 4x^2 + 3x - 2 \ \text{ and } 7x^9 + 8x^{-2} - 18x^{\frac{1}{3}} - 15x^{-\frac{2}{3}} + 7.$$

Is there anything missing when you differentiate an expression and then immediately integrate the result?
Why do you think this occurs?

Create some more of this kind for yourself.

(E)   What general conclusions can you make about the processes of differentiation and integration?

Do you need to refine your general result about the integral

$$\int x^n dx ?$$

(F)   You have probably deduced that the rule of integrating basic functions is closely related to differentiation.

Can you predict what the next results will be before you do them with DERIVE?

$$\int \cos(x)dx, \quad \int e^x \, dx, \quad \int \sin(x)dx, \quad \int e^{4x} \, dx,$$

$$\int e^{-0.2x} \, dx, \quad \int \cos(3x)dx, \quad \int \frac{1}{x}dx,$$

Summary

The results of DERIVE Activities 7A and 7B can now be brought together and are summarised in Table 7.3.

Table 7.3

Function f	Integral of f
$x^n \ (n \neq -1)$	$\dfrac{x^{n+1}}{(n+1)} + c$
$\dfrac{1}{x}$	$\ln(x) + c$
e^{ax}	$\dfrac{1}{a}e^{ax} + c$
$\sin(ax)$	$-\dfrac{1}{a}\cos(ax) + c$
$\cos(ax)$	$\dfrac{1}{a}\sin(ax) + c$

In each case c is an unknown constant called a *constant of integration*.

You have also seen that

(a)   the integral of a sum of functions is the sum of the integrals eg.

$$\int f(x) + g(x)dx = \int f(x)dx + \int g(x)dx \; .$$

(b)   If c is a constant then

$$\int cf(x)dx = c\int f(x)dx \; .$$

Notation

The process of finding areas under curves is called *definite integration*. It is a special summation process. We write the definite integral as

$$\int_a^b f(x)dx$$

where a and b are called *the limits* of the integration. It has a definite value.
Table 7.3 gives a set of rules for finding the integrals of the basic scientific functions. Formally we can write

$$\int f(x)dx = g(x) + c$$

where c is a constant called *the constant of integration* and $g(x)$ is a function such that

$$f(x) = \frac{dg}{dx} \; .$$

This is called the *indefinite integral* of f. It is a function of x unlike the definite integral which has a definite value. For example, if $f(x) = x$ then

$$\int x dx = \tfrac{1}{2}x^2 + c$$

whereas $\int_0^2 x dx = 2$ since it is the area of the triangle shown in Figure 7.10.

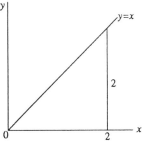

Figure 7.10

Fortunately we do not have to
evaluate areas when finding definite
integrals. Consider again the AREA
function introduced in DERIVE
Activity 7A.

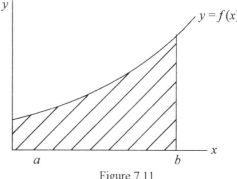

Figure 7.11

The function AREA($0,b,n$) was the area under the graph of $y = f(x)$ between
$x = 0$ and $x = b$ (see Figure 7.11).

Similarly AREA($0,a,n$) is the
area between $x = 0$ and $x = a$.

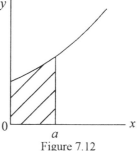

Figure 7.12

If we subtract the second area
from the first then we have the area
under the curve between $x = a$ and
$x = b$ (see Figure 7.13).

$$\int_a^b f(x)dx = \text{AREA}(0,b,n)$$
$$- \text{AREA}(0,a,n)$$

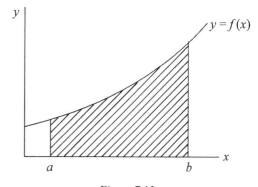

Figure 7.13

Now $\dfrac{d}{dx}\text{AREA}(0,x,n) = f(x)$ so the AREA function is the indefinite integral of f
(x).

$$\text{AREA}(0,x,n) = g(x) = \int f(x)dx$$

Substituting $x = b$ and $x = a$ gives the rule of evaluating definite integrals

$$\int_a^b f(x)dx = g(b) - g(a) .$$

We often use the notation $\int_a^b f(x)dx = [g(x)]_a^b = g(b) - g(a)$.
The following example shows the method of approach.

Example 7D

Evaluate $\int_{-1}^2 x^2 dx$.

Solution

From the table of standard integrals (Table 7.3)

$$\int x^2 dx = \frac{x^3}{3} + c$$

so that $g(x) = \frac{1}{3}x^3$ in this case. Hence

$$\int_{-1}^2 x^2 dx = g(2) - g(-1)$$

$$= \frac{2^3}{3} - \frac{(-1)^3}{3} = 3 .$$

Example 7E

Evaluate $\int_0^1 1 + \frac{x^2}{2} - \frac{x^4}{3} dx$.

Solution

$$\int_0^1 1 + \frac{x^2}{2} - \frac{x^4}{3} dx = \left[x + \frac{x^3}{6} - \frac{x^5}{15} \right]_0^1 .$$

Evaluating the function at $x = 0$ and $x = 1$ and subtracting gives

$$\int_0^1 1 + \frac{x^2}{2} - \frac{x^4}{3} dx = \left(1 + \frac{1}{6} - \frac{1}{15}\right) - (0) = \frac{33}{30} = 1.1 \ .$$

DERIVE can be used to evaluate definite integrals. The following commands will evaluate the integral in Example 7E.

Author $1 + \dfrac{x^2}{2} - \dfrac{x^4}{3}$ **Calculus Integrate**.

In response to CALCULUS INTEGRATE: Lower Limit   Upper Limit

type in 0 (Tab) 1 (enter). **Simplify** and **approX** gives 1.1 as required.

```
            2     4
           x     x
  1:   1 + --- - ---
           2      3

        1
        ⌠      2     4
        ⎮      x     x
  2:    ⎮ [1 + --- - ---] dx
        ⌡      2      3
        0

        11
  3:    --
        10

  4:   [1.1]
═══════════════════════════════════════════════════════════════
COMMAND: Author Build Calculus Declare Expand Factor Help Jump soLve Manage
         Options Plot Quit Remove Simplify Transfer moVe Window approX
Compute time: 0.0 seconds
Approx(3)                          Free:98%              Derive Algebra
```

Figure 7.14

Figure 7.14 shows the DERIVE screen for this activity. Notice that DERIVE shows the formal expression of the definite integral.

Exercise 7B

In each problem check your answers using DERIVE.

1.   Find the indefinite integrals of the following functions.

    (a) $4x^3$ (b) $3x^5$ (c) $x^{\frac{1}{2}}$

    (d) $13x^2 - 7x^3$ (e) $3x^5 + 2x^3 - x + 4$ (f) $6 + 3x - 2x^2$

    (g) $(4x+2)^2$ (h) $(1-x)^2$ (i) $\dfrac{1}{x^2}$

    (j) $x^{-0.7}$ (k) $1.7x^{-2.3}$ (l) $3x^{-1}$

    (m) $5x^{-6} + 3x^{-2}$ (n) $2x^{0.3} + \dfrac{1}{x}$ (o) $9x^{17} - 2x^4 + 3x^{-2}$

2.   Find the indefinite integrals of the following functions.

    (a) e^{2x} (b) e^{-5x} (c) $e^{0.1x}$

    (d) $3e^{4x}$ (e) $6e^{6x}$ (f) $-0.9e^{-0.5x}$

    (g) $4e^{3x} - 3e^{-2x}$ (h) $0.6e^{3.1x} - 0.9e^{-0.3x}$

3.   Find the indefinite integrals of the following functions.

    (a) $\sin(5x)$ (b) $\cos(1.5x)$ (c) $4\sin(3x)$

    (d) $3\sin(2x) - 2\cos(3x)$ (e) $2\sin(\pi x)$ (f) $1.5\cos(3\pi x)$

    (g) $2\sin(wx)$ where w is a constant

    (h) $1.5\cos(7x) + 0.3\sin(2x)$ (i) $e^{0.1x} + 2\sin(\pi x)$

    (j) $3x^2 + 4.2e^{-0.6x} + \cos(0.9x)$ (k) $\dfrac{1}{3x} + 0.5\sin(5x)$

4.   Evaluate the following integrals.

(a) $\int_1^2 x^2 dx$ (b) $\int_0^1 3x + 6 dx$ (c) $\int_{-1}^1 x^2 - 3x + 4 dx$

(d) $\int_0^1 2e^{3x} dx$ (e) $\int_1^3 e^x - e^{-2x} dx$ (f) $\int_{-1}^1 5e^{0.2x} dx$

(g) $\int_0^\pi \sin x dx$ (h) $\int_0^{\pi/2} \cos x dx$ (i) $\int_0^4 7e^{-0.1x} - 2x^{\frac{1}{2}} dx$

5.   Use integration to evaluate the area of the following regions:

(a)   between $y = 1 - x^2 + x^4$, $y = 0$, $x = 0$ and $x = 1$,

(b)   between $y = \sin x$ and the x-axis between $x = 0$ and $x = \pi$,

(c)   between $y = e^{-2x}$, $y = 0$, $x = 1$ and $x = 3$,

(d)   between $y = 2e^x$, $y = 1$, $x = 0$ and $x = 2$.

6.   Integrate the following.

(a) $(x-1)(x-2)$ (b) $x^2(x+2)$ (c) $\dfrac{1+x}{x^3}$

(d) $\dfrac{x^4 + 1}{x^2}$ (e) $\dfrac{a}{x^2} + b$ where a, b are constants

(f) $ax^2 + bx + c$ where a, b and c are constants.

7.   The gradient of a curve which passes through the point $(1,1)$ is given by $2 + 2x - x^2$. Find the equation of the curve.

8.   The gradient of a curve which passes through the point $(0,1)$ is given by $1 + x^2$. Find the equation of the curve.

9.   Evaluate the following integrals.

      (a) $\int t^2 dt$ (b) $\int 3t + 1 dt$ (c) $\int t^3 + 2t^2 dt$

      (d) $\int \dfrac{1}{t^2} dt$ (e) $\int 5t^7 + 4t^3 - t^{-1} dt$ (f) $\int (t-1)^2 dt$

      (g) $\int at^2 + bt + c\, dt$ where a, b, c are constants

      (h) $\int e^{2t}\, dt$ (i) $\int 2e^{0.1t}\, dt$ (j) $\int 2\sin(\pi t)\, dt$

      (k) $\int \dfrac{1}{t} dt$ (l) $\int 5\cos(0.1t)\, dt$ (m) $\int \sin(t) + \cos(t)\, dt$

      (n) $\int \dfrac{1}{v^2} dv$ (o) $\int p^{-\frac{1}{2}} dp$ (p) $\int u^2 + 3u + 8\, du$

      (q) $\int \dfrac{1}{w} dw$ (r) $\int e^{2u}\, du$ (s) $\int \sqrt{y}\, dy$

10.   The acceleration of an object is related to the velocity through the equation

$$a = \frac{dv}{dt}.$$

      (a)   If $a = t^2$ find v, given that $v = 2$ when $t = 0$.

      (b)   If $a = t + 1$ find v, given that $v = 1$ when $t = 1$.

      (c)   If $a = 3\cos(t)$ find v.

11.   The force on an object is related to the potential energy through the equation

$$F = \frac{dV}{dx}.$$

      (a)   Find the potential energy due to gravity if $F = mg$.

      (b)   Find the potential energy in an elastic string if $F = k(x-L)$.

7.3   MORE ABOUT AREAS

The introduction to integration has linked definite integrals to areas. However you must be careful in this interpretation. Consider the value of the following integral.

$$\int_{-1}^{1} x\,dx = \left[\frac{x^2}{2}\right]_{-1}^{1} = \frac{1}{2} - \frac{1}{2} = 0 \ .$$

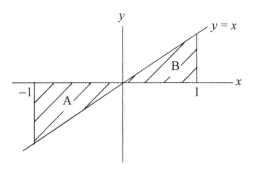

Figure 7.15

Does this mean that the area between the function $f(x) = x$ and the x-axis between $x = -1$ and $x = 1$ is zero. Figure 7.15 shows this area.
Clearly the area of triangle $A = \frac{1}{2}$ and the area of triangle $B = \frac{1}{2}$ so the total shaded area = 1. But the value of the integral is 0.
To understand this anomaly

consider the two integrals $\int_{-1}^{0} x\,dx$ and

$\int_{0}^{1} x\,dx$.

$$\int_{-1}^{0} x\,dx = \left[\frac{1}{2}x^2\right]_{-1}^{0} = 0 - \frac{1}{2} = -\frac{1}{2}$$

$$\int_{0}^{1} x\,dx = \left[\frac{1}{2}x^2\right]_{0}^{1} = \frac{1}{2} - 0 = \frac{1}{2}$$

When the function values are negative the integral of the function will be negative. Since areas are positive we can interpret the integral as an area by ignoring the sign of the integral. Hence

$$\text{Area } A = \left|\int_{-1}^{0} x\,dx\right| = \left|-\frac{1}{2}\right| = \frac{1}{2} \ .$$

It is important to interpret the value of an integral as an area with caution. It is advisable to sketch the graph of the function to be integrated and then integrate the function in portions to take account of the signs. Figure 7.16 shows the DERIVE screen for finding the area between $y = x^2 - 4$, $x = 0$, $x = 3$ and $y = 0$. Line #3 is the integral between $x = 0$ and $x = 2$. (Its value is negative because the graph is below the x-axis); and line #5 is the area between $x = 2$ and $x = 3$.

The required area is given by (line #5) + (–line #3) = 7.66666 (actually $7\frac{2}{3}$ by exact arithmetic).

1:    $x^2 - 4$

2:    $\displaystyle\int_0^2 (x^2 - 4)\ dx$

3:    $-\dfrac{16}{3}$

4:    $\displaystyle\int_2^3 (x^2 - 4)\ dx$

5:    $\dfrac{7}{3}$

COMMAND: **Author** Build Calculus Declare Expand Factor Help Jump soLve Manage
          Options Plot Quit Remove Simplify Transfer moVe Window approX
Enter option
Simp(4)   Free:99%   Derive Algebra

Figure 7.16

Exercise 7C

1.   Sketch the graph of $y = x^5$. Show that

$$\int_{-2}^{2} x^5 dx = 0 \ .$$

Use your sketch to explain your answer.

2.   Calculate the shaded areas in the following diagrams.

(a)

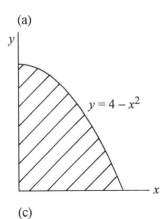

(b)

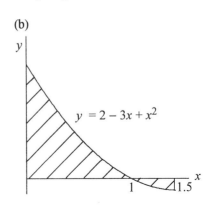

(c)

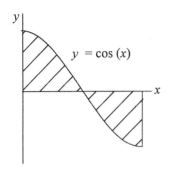

(d)

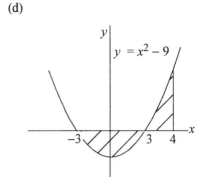

Figure 7.17

3.   Evaluate the following definite integrals and interpret the values in terms of areas.

(a) $\int_{-1}^{1} x - 2\,dx$ (b) $\int_{0.5}^{2} \frac{1}{x}\,dx$

(c) $\int_{0}^{2} x^2 - x\,dx$ (d) $\int_{-1}^{1} 2\mathrm{e}^{-x}\,dx$

(e) $\int_{-\pi/2}^{\pi/2} \sin(x)\,dx$ (f) $\int_{0}^{1} 3\mathrm{e}^{-x} - 2\,dx$

4.   In the following diagrams the graphs are symmetrical about either the x or y axes. The area of part of the region is given. Deduce the area of the whole shaded region and the value of the given integral.

(a)

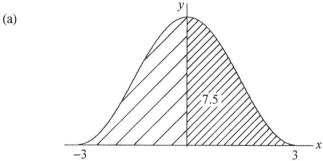

Evaluate $\int_{-3}^{3} f(x)\,dx$.

(b)

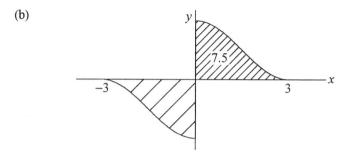

Evaluate $\int_{-3}^{3} f(x)\,dx$.

(c)

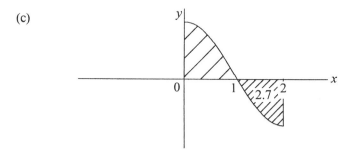

Evaluate $\int_{0}^{1} f(x)\,dx,\ \int_{1}^{2} f(x)\,dx,\ \int_{0}^{2} f(x)\,dx$.

Figure 7.18

7.4   INTEGRATION BY DIRECT SUBSTITUTION

The process of integration by substitution involves making a change of variable so that the given integral is transformed into a standard form. The technique is very powerful.

Example 7E

Evaluate $\int (3x-1)^4 \, dx$.

Solution

The substitution $u = 3x - 1$ will reduce $(3x-1)^4$ to u^4. However $\int u^4 dx$ involves two variables and so we must write dx in terms of du. We do this using the chain rule.

Now $u = 3x - 1$, so $\dfrac{du}{dx} = 3$ and hence $dx = \tfrac{1}{3} du$. So

$$\int (3x-1)^4 \, dx = \int u^4 \left(\tfrac{1}{3} du \right) = \tfrac{1}{3} \int u^4 du = \tfrac{1}{3} \left(\frac{u^5}{5} \right) + c$$

$$= \frac{1}{15} (3x-1)^5 + c .$$

Note that we write the answer in terms of x, not the changed variable u.

Example 7F

Evaluate $\int x(3x^2 + 2)^{-\frac{1}{2}} \, dx$.

Solution

The most difficult part of the function to be integrated is $(3x^2+2)^{-\frac{1}{2}}$. So we try $u = 3x^2 + 2$ as a substitution. So $\dfrac{du}{dx} = 6x$ and $xdx = \tfrac{1}{6} du$. Hence

$$\int x(3x^2 + 2)^{-\frac{1}{2}} \, dx = \int u^{-\frac{1}{2}} \left(\tfrac{1}{6} du \right) = \tfrac{1}{6} \int u^{-\frac{1}{2}} du = \tfrac{1}{6} (2u^{\frac{1}{2}}) + c$$

$$= \frac{1}{3} (3x^2 + 2)^{\frac{1}{2}} + c .$$

Example 7G

Evaluate $\int \dfrac{4x+1}{2x^2 + x + 3} dx$.

Solution

Again the function to be integrated suggests the substitution $u = 2x^2 + x + 3$. So

$\dfrac{du}{dx} = 4x + 1$ and $(4x+1)dx = du$. Hence

$$\int \frac{4x+1}{2x^2 + x + 3} dx = \int \frac{1}{u} du = \ln u + c$$
$$= \ln(2x^2 + x + 3) + c .$$

Example 7H

Evaluate $\int_0^{\pi/2} \sin(2t + \pi) dt$.

Solution

For definite integrals we change variables and the limits to obtain a new definite integral.

Choose the substitution $u = 2t + \pi$ so that $\dfrac{du}{dt} = 2$ and $dt = \frac{1}{2} du$.

Now compute the new limits. For the lower limit $t = 0$, $u = \pi$ and for the upper limit

$t = \dfrac{\pi}{2}$, $u = 2\pi$.

Hence changing variables

$$\int_0^{\pi/2} \sin(2t + \pi)\,dt = \int_\pi^{2\pi} \sin u(\tfrac{1}{2}\,du) = \frac{1}{2}\int_\pi^{2\pi} \sin u\,du$$

$$= \frac{1}{2}[-\cos u]_\pi^{2\pi}$$

$$= \frac{1}{2}(-\cos 2\pi + \cos \pi) = -1.$$

All of these integrals can be evaluated using DERIVE.

Figure 7.19 shows the DERIVE screen for these examples.

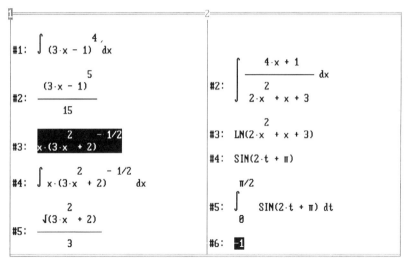

Figure 7.19

Exercise 7D

Check your answers using DERIVE.

1.. Evaluate the following indefinite integrals.

(a) $\int x(4x^2 - 1)dx$ (b) $\int \sqrt{x-1}dx$

(c) $\int \sin(2x+1)dx$ (d) $\int \dfrac{2}{3x+2}dx$

(e) $\int \dfrac{x}{5x^2 - 2}dx$ (f) $\int e^{2x+1} dx$

(g) $\int (4x+1)^3 dx$ (h) $\int \dfrac{1}{(1+x)^2}dx$

(i) $\int \dfrac{1}{5x-3}dx$ (j) $\int xe^{x^2} dx$

(k) $\int (1+x)^{0.3} dx$ (l) $\int \tan x dx$ (let $u = \cos x$)

(m) $\int (2x+1)(x^2 + x + 3)^{0.5} dx$ (n) $\int \dfrac{2x+3}{x^2+3x-5}dx$

(o) $\int (4t - 11)^5 dt$ (p) $\int \dfrac{1}{(3-2v)^2}dv$

(q) $\int \cos(2y+\pi)dy$ (r) $\int \dfrac{t}{1+t^2}dt$

(s) $\int \cos^2 \theta \sin\theta d\theta$ (t) $\int \sin^4 (2\theta)\cos(2\theta)d\theta$

2. Evaluate the following definite integrals.

(a) $\int_0^{\pi/2} \sin(3x - \pi)dx$ (b) $\int_1^3 \dfrac{1}{5x-3}dx$

(c) $\int_1^2 \dfrac{1}{2t+1}dt$ (d) $\int_1^3 \sqrt{2x-1}\, dx$

(e) $\int_0^2 \dfrac{t^2}{1+t^3}dt$ (f) $\int_0^1 xe^{x^2} dx$

(g) $\int_{-1}^2 e^{3u-5}\, du$ (h) $\int_{-1}^1 (4v+1)^3 dv$

## 7.5	INTEGRATION BY INDIRECT SUBSTITUTIONS

The substitutions in section 7.4 are called *direct substitutions* because they tend to be fairly obvious from the function to be evaluated. Sometimes it is necessary to make trigonometric substitutions which may seem less than obvious. These are called *indirect substitutions*.

Example 7I

Evaluate $\int \dfrac{1}{\sqrt{16-x^2}} dx$.

Solution

Let $x = 4\sin\theta$ then $\dfrac{dx}{d\theta} = 4\cos\theta$ so $dx = 4\cos\theta d\theta$. Hence

$$\int \frac{1}{\sqrt{16-x^2}} dx = \int \frac{1}{\sqrt{16-16\sin^2\theta}} (4\cos\theta d\theta)$$

$$= \int \frac{1}{4\sqrt{1-\sin^2\theta}} (4\cos\theta d\theta) = \int \frac{1}{(4\cos\theta)} (4\cos\theta d\theta)$$

since $1 - \sin^2\theta = \cos^2\theta$

$$= \int 1 d\theta = \theta + c$$

$$= \arcsin\left(\frac{x}{4}\right) + c .$$

Example 7J

Evaluate $\int \dfrac{1}{1+x^2} dx$.

Solution

Let $x = \tan\theta$ then $\dfrac{dx}{d\theta} = \sec^2\theta$ so $dx = \sec^2\theta d\theta$. Hence

$$\int \frac{1}{1+x^2} dx = \int \frac{1}{(1+\tan^2\theta)} (\sec^2\theta d\theta)$$

$$= \int \frac{1}{(\sec^2\theta)} (\sec^2\theta d\theta) \quad \text{since } 1+\tan^2\theta = \sec^2\theta$$

$$= \int 1 d\theta = \theta + c$$

$$= \arctan x + c \ .$$

As a general rule if the function to be integrated involves $a + bx^2$ then try

$x = \sqrt{\dfrac{a}{b}} \tan\theta$; and if the function to be integrated involves $\sqrt{a - bx^2}$ then try

$x = \sqrt{\dfrac{a}{b}} \sin\theta$. In the following exercise the substitutions are given.

Exercise 7E

Evaluate the following integrals using the suggested substitution. Check your answers using DERIVE.

(a) $\int \dfrac{1}{\sqrt{1-4x^2}} dx \qquad x = \tfrac{1}{2}\sin\theta$

(b) $\int \dfrac{1}{1+4t^2} dt \qquad t = \tfrac{1}{2}\tan\theta$

(c) $\int \dfrac{1}{\sqrt{9-x^2}} dx \qquad x = 3\sin\theta$

(d) $\int \dfrac{1}{1+9x^2} dx \qquad x = \tfrac{1}{3}\tan\theta$

(e) $\int_0^{\sqrt{3}/2} \dfrac{1}{1+4u^2} du \qquad u = \tfrac{1}{2}\tan\theta$

(f) $\int_1^{\sqrt{2}} \dfrac{1}{\sqrt{2-t^2}} dt \qquad t = \sqrt{2}\sin\theta$

7.6   A PAIR OF TRIGONOMETRIC INTEGRALS

There are two functions $\sin^2 x$ and $\cos^2 x$ whose integrals involve techniques which do not fit into the classification of substitutions.

These integrals are $\int \sin^2 x\, dx$ and $\int \cos^2 x\, dx$.

For these functions we use the double angle formulas

$$\cos 2x = 1 - 2\sin^2 x \quad \text{and} \quad \cos 2x = 2\cos^2 x - 1 .$$

$$\int \sin^2 x\, dx = \int \tfrac{1}{2}(1 - \cos 2x)\, dx = \frac{1}{2}x - \frac{1}{4}\sin 2x + c$$

$$\int \cos^2 x\, dx = \int \tfrac{1}{2}(1 + \cos 2x)\, dx = \frac{1}{2}x + \frac{1}{4}\sin 2x + c$$

7.7   INTEGRATION BY PARTS

In this section we introduce a rule for integrating products of functions. As with all rules in integration the method is not always guaranteed to be applicable in every case! We begin with the rule for differentiating a product

$$\frac{d}{dx}(uv) = u\frac{dv}{dx} + v\frac{du}{dx} .$$

Integrating both sides with respect to x

$$\int \left[\frac{d}{dx}(uv)\right] dx = \int u\frac{dv}{dx}\, dx + \int v\frac{du}{dx}\, dx$$

so

$$uv = \int u\frac{dv}{dx}\, dx + \int v\frac{du}{dx}\, dx .$$

Rearranging gives

$$\int u \frac{dv}{dx} dx = uv - \int v \frac{du}{dx} dx .$$

This rule can be used to integrate some products of functions. It changes the integral on the left hand side to a product of two functions and another integral on the right hand side. The method is called *integration by parts* and the aim is to make the new integral easier than the original one. The following example shows how the method works.

Example 7I

Evaluate $\int x \cos x \, dx$.

Solution

We have to integrate the product of two functions x and $\cos x$.

Suppose that we let $u = x$ and $\frac{dv}{dx} = \cos x$.

Then $v = \sin x$ and $\frac{du}{dx} = 1$. So

$$\int u \frac{dv}{dx} dx = uv - \int v \frac{du}{dx} dx \text{ becomes}$$

$$\int x \cos x dx = x \sin x - \int (\sin x)(1) dx .$$

The new integral is quite straightforward. Hence

$$\int x \cos x dx = x \sin x + \cos x + c .$$

In evaluating the integral we made the choice $u = x$. Suppose we had let

$u = \cos x$ and $\dfrac{dv}{dx} = x$. Then $v = \frac{1}{2}x^2$ and $\dfrac{du}{dx} = -\sin x$. So

$$\int x \cos x\, dx = \frac{1}{2}x^2 \cos x - \int \frac{1}{2}x^2 (-\sin x)\, dx$$

$$= \frac{1}{2}x^2 \cos x + \frac{1}{2}\int x^2 \sin x\, dx .$$

Clearly the new integral is harder than the original one. So the substitution for u and

$\dfrac{dv}{dx}$ needs to be made carefully.

Example 7J

Evaluate $\int x^2\, e^{-3x}\, dx$.

Solution

The choice here is $u = x^2$ because after differentiating x^2 twice this will give a constant.

Let $u = x^2$ and $\dfrac{dv}{dx} = e^{-3x}$. Then $\dfrac{du}{dx} = 2x$ and $v = -\frac{1}{3}e^{-3x}$.

$$\int u \frac{dv}{dx}dx = uv - \int v \frac{du}{dx}dx \quad \text{becomes}$$

$$\int x^2\, e^{-3x}\, dx = -\frac{1}{3}x^2\, e^{-3x} - \int\left(-\frac{1}{3}e^{-3x}\right)(2x)\, dx$$

$$= -\frac{1}{3}x^2\, e^{-3x} + \frac{2}{3}\int x e^{-3x}\, dx . \qquad *$$

The new integral must be evaluated in parts.

Let $u = x$ and $\dfrac{dv}{dx} = e^{-3x}$.

Then $\dfrac{du}{dx} = 1$ and $v = -\tfrac{1}{3}e^{-3x}$. So

$$\int x e^{-3x}\, dx = -\tfrac{1}{3} x e^{-3x} - \int \left(-\tfrac{1}{3}e^{-3x}\right)(1)\,dx$$

$$= -\tfrac{1}{3} x e^{-3x} + \tfrac{1}{3}\int e^{-3x}\, dx$$

$$= -\tfrac{1}{3} x e^{-3x} - \tfrac{1}{9} e^{-3x} + c \ .$$

Substituting into * gives

$$\int x^2 e^{-3x}\, dx = -\frac{1}{3} x^2 e^{-3x} + \frac{2}{3}\left[-\frac{1}{3} x e^{-3x} - \frac{1}{9} e^{-3x} + c\right]$$

$$= -\left(\frac{1}{3} x^2 + \frac{2}{9} x + \frac{2}{27}\right) e^{-3x} + A$$

where A is an arbitrary constant.

Exercise 7F

Check your answers using DERIVE.

1.   Evaluate the following indefinite integrals.

(a) $\int x \sin x\, dx$ (b) $\int x e^x\, dx$

(c) $\int x^2 e^{2x}\, dx$ (d) $\int x \sin 3x\, dx$

(e) $\int t^2 e^t\, dt$ (f) $\int x \ln x\, dx$

(g) $\int \ln x\, dx$ [Hint: let $u = \ln x$ and $\dfrac{dv}{dx} = 1$].

2.   Evaluate the following definite integrals.

(a) $\int_0^1 u\,e^u\,du$ (b) $\int_0^{\pi/2} x\cos 2x\,dx$

(c) $\int_2^4 x\sqrt{x-1}\,dx$ (d) $\int_0^2 t^2\,e^{-2t}$

(e) $\int_0^{\pi/2} e^x \sin x\,dx$ (f) $\int_0^{\pi/3} e^{-2x} \sin 3x\,dx$

7.8   USE OF PARTIAL FRACTIONS

The integral of the form $\int \dfrac{1}{a+bx}\,dx$ can be evaluated to give $\dfrac{1}{b}\ln(a+bx)$ and this provides a standard form for evaluating integrals of functions given as $f(x)/g(x)$ where $f(x)$ and $g(x)$ are polynomials. The technique is to transform the integral using a partial fraction expansion. The following example illustrates the approach.

Example 7K

Evaluate $\int \dfrac{x+1}{x^2+x-2}\,dx$.

Solution

The factors of $x^2 + x - 2$ are $(x+2)$ and $(x-1)$, so we write the function in the form

$$\frac{x+1}{x^2+x-2} = \frac{x+1}{(x+2)(x-1)} = \frac{A}{(x+2)} + \frac{B}{(x-1)}\;.$$

This is called the *partial fraction* form.

If we combine the two *partial fractions* on the right hand side we get

$$\frac{A}{(x+2)} + \frac{B}{(x-1)} = \frac{A(x-1)+B(x+2)}{(x+2)(x-1)}\;.$$

Comparing the numerators of the first and last fractions leads to

$$x + 1 = A(x-1) + B(x+2)\;.$$

To find the values of A and B first, let $x = 1$ to give $2 = 0 + 3B$, so that $B = \frac{2}{3}$. Then let $x = -2$ to give $-1 = -3A + 0$, so that $A = \frac{1}{3}$.

Alternatively we could compare coefficients of x on each side of the equation. Since

$$x+1 = (A+B)x + 2B - A$$

then

$$A + B = 1 \quad \text{and} \quad 2B - A = 1 .$$

Solving these gives $A = \frac{1}{3}$ and $B = \frac{2}{3}$ as before.

Now the integral can be written as

$$\int \frac{x+1}{x^2 + x - 2} dx = \int \frac{\frac{1}{3}}{(x+2)} + \frac{\frac{2}{3}}{(x-1)} dx$$

$$= \frac{1}{3}\ln(x+2) + \frac{2}{3}\ln(x-1) + c .$$

This method of approach can be applied provided the denominator $g(x)$ has real linear factors. However there are many polynomials with quadratic factors $ax^2 + bx + c$ which cannot be factorized into two linear factors. This then requires evaluation of integrals of the form

$$\int \frac{1}{ax^2 + bx + c} dx .$$

Example 7L

Evaluate $\int \dfrac{1}{x^2 - 4x + 6}\,dx$.

Solution

The first step is to consider the function $x^2 - 4x + 6$ and complete the square

$$x^2 - 4x + 6 = (x - 2)^2 + 2 \ .$$

Now let $u = x - 2$ so that $dx = du$. On substitution the integral becomes

$$\int \frac{1}{x^2 - 4x + 6}\,dx = \int \frac{1}{(x-2)^2 + 2}\,dx$$

$$= \int \frac{1}{u^2 + 2}\,du \ .$$

Now with $u = \sqrt{2}\,\tan\theta$ and $du = \sqrt{2}\,\sec^2\theta\,d\theta$ this integral becomes

$$\int \frac{1}{u^2 + 2}\,du = \int \frac{1}{2(1 + \tan^2\theta)}\sqrt{2}\,\sec^2\theta\,d\theta$$

$$= \int \frac{1}{\sqrt{2}}\,d\theta = \frac{\theta}{\sqrt{2}} + c \ .$$

Replacing θ by $\arctan(u/\sqrt{2})$ and u by $x - 2$ gives

$$\int \frac{1}{x^2 - 4x + 6}\,dx = \frac{1}{\sqrt{2}}\arctan\!\left((x-2)/\sqrt{2}\right) + c \ .$$

Exercise 7G

Evaluate the following integrals. Check your results using DERIVE.

(a) $\int \dfrac{2}{(x+1)(x-1)} dx$

(b) $\int \dfrac{1}{(2x-1)(x+2)} dx$

(c) $\int \dfrac{5}{(x+1)(x-3)} dx$

(d) $\int \dfrac{3}{(t-2)(t-3)} dt$

(e) $\int \dfrac{2v+1}{(2v-1)(2v+3)} dv$

(f) $\int_1^2 \dfrac{2x+1}{(x+2)(x+3)} dx$

(g) $\int \dfrac{2x^2-9x+1}{x(x-1)(x+3)} dx$

(h) $\int_0^1 \dfrac{t(t-2)}{(t+1)(t^2+1)} dt$

(i) $\int \dfrac{1}{x^2-9x+25} dx$

(j) $\int_0^1 \dfrac{1}{(x+1)^2(x+2)} dx$

Exercise 7H

Integration is a vast topic which involves several different techniques. The skill is deciding which technique to use. The following integrals will give you practice at choosing the right method. Check your answers using DERIVE.

1.   Evaluate the following integrals.

(a) $\int \sin x \cos x\, dx$

(b) $\int \dfrac{3x^2}{x^3-3} dx$

(c) $\int e^{2x} - \sin 4x\, dx$

(d) $\int_1^4 \dfrac{dx}{\sqrt{5-x}}$

(e) $\int x \sin^2 x\, dx$

(f) $\int_0^1 \dfrac{1}{4+t^2} dt$

(g) $\int_0^1 \dfrac{1}{4-t^2} dt$

(h) $\int \dfrac{t}{4+t^2} dt$

(i) $\int_0^1 t\, e^{4t^2}\, dt$

(j) $\int \dfrac{5}{t^2+t-6} dt$

(k) $\int \dfrac{u^2+2}{u} du$

(l) $\int \sqrt{1-v^2}\, dv$

(m) $\int u^2(u^2+1)\, du$

(n) $\int_0^2 e^{(4t-1)}\, dt$

(o) $\int u^2 \sin u\, du$

(p) $\int t\, e^{3t}\, dt$

2. The speed of an object travelling in a straight line is given by $(3t+5)^2$ ms^{-1} where t is the time. Find the formula for the distance travelled in the first 5 seconds.

3. In a physics experiment it is found that the rate of change of temperature is inversely proportional to $(2t+3)^2$. Find a formula for the temperature as a function of time.

4. The force acting on an object due to an elastic spring is given by

   $F(x) = 15x - 20$. The potential energy $V(x)$ is related to the force by $\dfrac{dV}{dx} = F$. Use integration to find a formula for $V(x)$.

5. The gradient of a curve at any point is given by $\dfrac{dy}{dx} = \dfrac{1}{x^2 - 3x + 2}$. If the curve passes through the point $(0,1)$ find the equation of the curve.

6. Find the total area enclosed by the curve $y = x(x-1)(x-2)$ and the x-axis between $x = 0$ and $x = 4$.

7. Find the area enclosed by the line $y = 2$ and the curve $y = x(3-x)$.

8. A function $f(t)$ is given in tabular form as

t	$f(t)$
1.8	6.050
2.0	7.389
2.2	9.025
2.4	11.023
2.6	13.464
2.8	16.445
3.0	20.086
3.2	24.533
3.4	29.964

   Find an approximate value for $\int_{1.8}^{3.4} f(t)dt$.

9.   Consider the function $f(x) = 2x - 3$.

    (a)   Draw a graph of the function and find the area between $f(x)$, the x-axis, $x = 1$ and $x = 2$.

    (b)   Evaluate $\int_1^2 (2x - 3)\,dx$.

    Are your answers to (a) and (b) equal? Does an integral always equal an area?

10.   An oil droplet is falling through a medium which produces a resistance proportional to $v^{3/2}$ where v is its downward velocity. The time taken for it to reach a velocity of $u/2$, where u is its terminal velocity, is given by

$$t = \frac{u}{g} \int_0^{0.5} \frac{1}{1 - x^{3/2}}\,dx \ .$$

Evaluate this expression using DERIVE.

11.   The fraction of internally radiated heat energy falling on a thermocouple which is centrally placed on the axis of a hot tube of length L and radius R is given by

$$H = \frac{1}{2} \int_{-1}^{1} \frac{a}{\left[1 + (ax)^2 \right]^{3/2}}\,dx$$

where $a = \dfrac{L}{2R}$. Find the value of H for a tube of length 8 cm and radius 8 cm.

8

Numerical methods

8.1 NEWTON-RAPHSON METHOD

In Chapter 5 you will have seen how to use iterative formulae to find the solutions of equations. In this section a method for formulating iterative formulae that converge will be developed to solve equations of the form $f(x) = 0$.

DERIVE ACTIVITY 8A

(A)   Consider the equation

$$x^3 + 2x - 2 = 0 .$$

   (i)   **Author** and **Plot** the graph of $y = x^3 + 2x - 2$.

   (ii)   Use the **Transfer** command to **Load** the **Utility** file DIF_APPS. This investigation shows how to improve the accuracy of solutions by using the TANGENT command. Take $x = 1$ as a first estimate of the solution of $x^3 + 2x - 2 = 0$.
   Author and **Simplify** TANGENT ($x^3 + 2x - 2$, x, 1). Now **Plot** the tangent to the curve at $x = 1$. Note that the tangent crosses the x-axis at a point closer to the curve than the original estimate of $x = 1$. Use the **soLve** command on the equation of the tangent to find the improved estimate of x the solution of the equation which should be 0.8.

   (iii) Move the cursor to (0.8,0) and **Centre**. Use **Scale** to select the scales x:0.05 and y:0.1. Find the equation of the tangent to the curve at the new estimate of the solution, in this case $x = 0.8$. Plot this tangent to the curve. Also find the improved estimate of the solution of the equation, by using **soLve** on the equation of the tangent.

(B)   Repeat the activity for the equation

$$x^4 - 1.4x^3 - 1.8x^2 + x = 0.$$

(C)   By drawing a tangent to the curve at the first estimate given below, find an improved estimate of the solution. In each case use the centre and scale suggested.

(i) $\dfrac{x^4}{10} - 1 = 0$, first estimate $x = 2$.

(Centre at $(1.5,0)$, scale: x: 0.5 y: 0.5).

(ii) $\cos x - x = 0$, first estimate $x = 1$.

(Centre at $(0,0)$, scale: x: 1, y: 1).

(iii) $e^x - x^2 = 0$, first estimate $x = -1$.

(Centre at $(0,0)$, scale: x: 1, y: 1).

From DERIVE activity 8A you will have seen how by repeatedly drawing tangents it is possible to move closer and closer to the solution. You may also have noticed that only a few steps are needed to produce a good result compared with the methods used in Chapter 5.

Figure 8.1 shows the graph of a function $f(x)$, and how the tangent to the curve gives an improved estimate x_{n+1} from a previous estimate x_n.

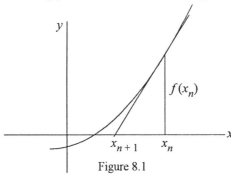

Figure 8.1

The gradient of the tangent at $x = x_n$ is given by $f'(x_n)$. It is also possible to find the gradient of the tangent by examining the triangle shown in Figure 8.1. The height of the curve at x_n is $f(x_n)$, so the slope of the tangent is given by

$$\frac{f(x_n)}{x_n - x_{n+1}} .$$

These two expressions can now be equated to give

$$f'(x_n) = \frac{f(x_n)}{x_n - x_{n+1}} .$$

This can be rearranged to give

$$x_n - x_{n+1} = \frac{f(x_n)}{f'(x_n)}$$

or

$$x_{n+1} = x_n - \frac{f(x_n)}{f'(x_n)} .$$

This iterative scheme is known as the *Newton-Raphson method* and provides a powerful method for finding the solutions of equations.

Example 8A

Find the positive solution of $x^4 - x^2 - 1 = 0$.

Solution

Before the Newton-Raphson method can be applied a first estimate of the solution needs to be found. A sign-search procedure can be used to do this. Table 8.1 below shows this.

Table 8.1

x	0	1	2
$x^4 - x^2 - 1$	-1	-1	11

A solution of $x^4 - x^2 - 1 = 0$ lies between $x = 1$ and $x = 2$. As the sign changes between $x = 1$ and $x = 2$, it is sensible to choose a value in this range as the first estimate. In this case we choose $x_1 = 1$.

Here $f(x)$ is defined as

$$f(x) = x^4 - x^2 - 1$$

and differentiating gives

$$f'(x) = 4x^3 - 2x .$$

So the Newton-Raphson method

$$x_{n+1} = x_n - \frac{f(x_n)}{f'(x_n)}$$

gives the iterative formula

$$x_{n+1} = x_n - \frac{(x_n^4 - x_n^2 - 1)}{(4x_n^3 - 2x_n)} .$$

Taking $x_1 = 1$ we can calculate x_2, x_3, ...

$$x_2 = 1 - \frac{(1^4 - 1^2 - 1)}{(4 \times 1^3 - 2 \times 1)} = 1.5$$

$$x_3 = 1.5 - \frac{(1.5^4 - 1.5^2 - 1)}{(4 \times 1.5^3 - 2 \times 1.5)} = 1.327 .$$

Repeating the process leads to the values

$$x_4 = 1.276$$
$$x_5 = 1.272$$
$$x_6 = 1.272$$

The positive solution of the equation $x^4 - x^2 - 1 = 0$ is 1.272 to 3 decimal places.

DERIVE ACTIVITY 8B

(A)   Use DERIVE to solve the equation $x^2 = e^x$ by following the procedure below.

(i)   Rewrite $x^2 = e^x$, in the form $x^2 - e^x = 0$, so that $f(x) = x^2 - e^x$.

(ii)   **Author** and **Plot** this function. Read off a first solution of the equation.

(iii)   Differentiate this function.

(iv)   Use the F3 button to assist you to **Author** the Newton-Raphson formula

$$x - \frac{(x^2 - e^x)}{(2x - e^x)} \; .$$

(v)   Use **Manage**, **Substitute** and **approX** to find an improved estimate using the first estimate taken from the graph in (ii) and the Newton-Raphson formula.

(vi)   Repeat (iv) until you can give the solution correct to 6 decimal places.

(vii)   Now **Author** (using the F3 button)

$$\text{ITERATES}\left(x - \frac{(x^2 - e^x)}{(2x - e^x)}, x, -1 \right)$$

and then **approX** this expression. The ITERATES command repeatedly applies the Newton-Raphson formula specified with a starting point of -1. Notice how quickly the method converges to a solution. Verify that this solution satisfies the original equation.

(viii) Try using other starting points. Do they all converge to the same value?

(B)   Consider the equation $x^3 - 5x^2 + 4x + 2 = 0$.

   (i)   Set up the Newton-Raphson formula for this equation in DERIVE.

   (ii)   Find solutions by using ITERATES and **approX**, for each of the
          following first estimates.

          (a) 0   (b)   1   (c)   5   (d)   2.7

   (iii) **Plot** $x^3 - 5x^2 + 4x + 2$ and check that the 3 solutions you obtained in (ii)
          are reasonable.

   (iv)   It seems strange that by starting at 2.7 you obtain the negative solution.
          To see why this happens, **Transfer**, **Load** the **Utility** file DIF_APPS.
          Use the TANGENT command to find the equation of the tangent to the
          curve at $x = 2.7$. Plot this tangent and explain why it leads to the
          negative solution.

(C)   Set up the Newton-Raphson formula in DERIVE to solve the following
       equation.

$$x^3 - 3x - 5 = 0$$

       Plot the graph of the appropriate function $f(x)$. Calculate the values of x for
       which $f'(x) = 0$. Draw the tangent to the graph at these values of x. Explain
       why these values should be avoided as first estimates in the Newton-Raphson
       formula.

(D)   Repeat (C) for

   (i) $x^2 - 2\sin x = 0$
   (ii) $x - e^{x^2} = 0$

       In each case give ranges of values of x for first estimates to use in the Newton-
       Raphson formula.

Summary

The Newton Raphson iterative formula converges to a limit very quickly for most initial values. But there can be problems! Particular care is needed in the choice of the first guess x_1. Consider the function in figure 8.2.

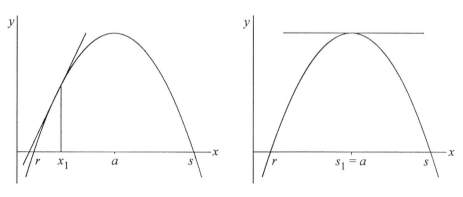

Figure 8.2   Figure 8.3

Suppose we wish to find the solution at $x = r$ and there is a local maximum at $x = a$. If we choose our starting iterate x_1 between r and a then the Newton-Raphson method will converge to $x = r$. However if we choose x_1 between a and s the method will converge to $x = s$ and not the root that we require.

Clearly if we choose $x_1 = a$, the turning point, then we are in trouble because the tangent is parallel to the x-axis (see figure 8.3). Finally if $\dfrac{df}{dx} = 0$ at the root then the Newton-Raphson method is not an appropriate method to use.

DERIVE has a built in function for solving non-linear equations, it is called **NEWTONS** and can be found in the Utility file **SOLVE.MTH**. The function f in the equation $f(x) = 0$, the variable x, the first iterate x_1 are entered as vectors. For example, to solve $x - e^{-x} = 0$ we **Author**

$$\text{NEWTONS}\big([x - e^{-x}], [x], [1], 6\big)$$

and **approX** to give a column of six iterates. Try this function to check your answers in Exercise 8A.

Exercise 8A

1. Use the Newton-Raphson method to solve the equations

    (a) $x - 2\sin x = 0$ (b) $x^3 - x + 8 = 0$

    (c) $x^3 = e^x$ (d) $\cos x = x$

    (e) $x^4 = x + 2$ (f) $\dfrac{1}{x} = e^x$

2. By writing $x^n = a$, find an iterative formula to find the nth root of a number. Use it to find

    (a) $\sqrt[3]{10}$ (b) $\sqrt[5]{8}$ (c) $\sqrt[10]{100}$.

3. Consider the equation $x^3 - 3x = 0$.

    (a)   Sketch or **Plot** the curve $y = x^3 - 3x$.

    (b)   There are two values of x for which the Newton-Raphson formula does not converge. Identify these points by drawing tangents and then explain why the method fails.

4. Find the positions of the turning points of the curve $y = e^x - 4x^2$.

5. A motor under load generates heat at a constant rate and radiates it at a rate proportional to the excess temperature T, so that at time t (in minutes)

    $$T = \frac{10}{k} - \frac{10}{k}e^{-kt}$$

    where k is a positive constant. After 10 minutes the temperature rise is 50°C. Find the value of k to four decimal places using the Newton-Raphson method.

8.2   APPROXIMATING FUNCTIONS BY A SERIES

DERIVE ACTIVITY 8C

(A)   (i)   **Author** and **Plot** the function $\sin x$. Then **Author** and **Plot** the series

$$x - \frac{x^3}{6} + \frac{x^5}{120} - \frac{x^7}{5040} \; .$$

(ii) Compare the two graphs. What properties do they have in common? For what range of values does the series provide a good approximation to $\sin x$?

(B)   (i)   **Author** and **Plot** the function e^x. Then **Author** and **Plot** the series

$$1 + x + \frac{x^2}{2} + \frac{x^3}{6} + \frac{x^4}{24} \; .$$

(ii)   Again compare the two graphs. What features do they have in common? For what range of values does the series provide a good approximation to e^x?

(iii)   Now use the **Manage, Substitute** commands to replace x by $2x$. Plot the series that you obtain. What function do you think that this might represent? **Author** and **Plot** your ideas to check them.

(C)   Each of the series below approximate one of the functions A, B and C.

A $\dfrac{1}{x}$ B $\cos(x)$ C $\sin(2x)$

(i) $1 - \dfrac{x^2}{2} + \dfrac{x^4}{24} - \dfrac{x^6}{720}$ (ii) $6 - 15x + 20x^2 - 15x^3 + 6x^4 - x^5$

(iii) $2x - \dfrac{4x^3}{3} + \dfrac{4x^5}{15} - \dfrac{8x^7}{315}$

Author and **Plot** each series. Then try to identify the function. **Author** and **Plot** the function to check your prediction.

Over what range of values is each approximation reasonable?

8.3   THE MACLAURIN SERIES

Some of the series considered above such as e^x, $\sin x$ and $\cos x$ gave their best approximations when x was close to zero, but as x increased or decreased, the goodness of the approximation would decrease. Such approximations are known as *Maclaurin series* and are based on information about the function at $x = 0$.

To form such a series consider an approximation of the form of a polynomial

$$f(x) = a_0 + a_1 x + a_2 x^2 + a_3 x^3 + a_4 x^4 + \cdots$$

where a_i are coefficients that have to be determined to give a good approximation.

When $x = 0$, then $f(0) = a_0$ as all the other terms will be zero.

Differentiating the series four times gives

$$f'(x) = a_1 + 2a_2 x + 3a_3 x^2 + 4a_4 x^3 + \cdots$$
$$f''(x) = 2a_2 + 6a_3 x + 12a_4 x^2 + \cdots$$
$$f'''(x) = 6a_3 + 24a_4 x + \cdots\cdots$$
$$f^{iv}(x) = 24a_4 + \cdots\cdots\cdots .$$

Now substituting $x = 0$ into each of these expressions gives

$$f'(0) = a_1$$
$$f''(0) = 2a_2$$
$$f'''(0) = 6a_3$$
$$f^{iv}(0) = 24a_4$$

So the coefficients a_i are determined by the values of $f(x)$ and its derivatives when $x = 0$.

$$a_0 = f(0) \quad a_1 = f'(0) \quad a_2 = \frac{f''(0)}{2} \quad a_3 = \frac{f'''(0)}{6} \quad a_4 = \frac{f^{iv}(0)}{24} .$$

In general

$$a_n = \frac{f^{(n)}(0)}{n!}.$$

So the Maclaurin series can be written as

$$f(x) = f(0) + f'(0)x + \frac{f''(0)x^2}{2!} + \frac{f'''(0)x^3}{3!} + \cdots + \frac{f^n(0)x^n}{n!} + \cdots.$$

This is in fact a series with an infinite number of terms, but in practice only a few of the first terms are used to produce the approximations. For this reason the approximation is only valid for values of x close to 0.

Example 8B

Find the Maclaurin series for sin x, giving the first 3 non-zero terms of the series.

Solution

The first step is to find the derivatives of sin x and evaluate each of these and sin x at $x = 0$, as shown below.

$$
\begin{array}{ll}
f(x) = \sin x & f(0) = 0 \\
f'(x) = \cos x & f'(0) = 1 \\
f''(x) = -\sin x & f''(0) = 0 \\
f'''(x) = -\cos x & f'''(0) = -1 \\
f^{iv}(x) = \sin x & f^{iv}(0) = 0 \\
f^{v}(x) = \cos x & f^{v}(0) = 1
\end{array}
$$

As there are now 3 non-zero derivatives it is possible to form the Maclaurin series. The series is defined as

$$f(x) = f(0) + f'(0)x + \frac{f''(0)x^2}{2!} + \frac{f'''(0)x^3}{3} + \frac{f^{iv}(0)x^4}{4!} + \frac{f^{v}(0)x^5}{5!} + \cdots$$

In this case we obtain

$$f(x) = 0 + 1 \times x + \frac{0 \times x^2}{2!} + \frac{(-1) \times x^3}{3!} + \frac{0 \times x^4}{4!} + \frac{1 \times x^5}{5!} + \cdots$$

i.e.

$$\sin x = x - \frac{x^3}{6} + \frac{x^5}{120} + \cdots \cdots.$$

Exercise 8B

1.   Find the Maclaurin series of the following functions giving the first three non-zero terms.

   (a) e^x

   (b) $\cos x$

   (c) $\sin(x^2)$

   (d) $\dfrac{1}{1+x}$

   (e) $\tan x$

   (f) $\sqrt{1-x}$

   (g) $\ln(\cos x)$

   (h) e^{x^2}

   Use DERIVE to compare each function with its Maclaurin series and in each case suggest the range of values of x that the approximation is a good one.

   For parts (d) and (f) compare your Maclaurin series with the series obtained using the binomial theorem.

2.   Use the Maclaurin series expansion of $(1+x)^n$ to obtain the binomial theorem

$$(1+x)^n = 1 + nx + \frac{n(n-1)x^2}{2!} + \frac{n(n-1)(n-2)x^3}{3!} + \cdots$$

   Use DERIVE to plot $(1+x)^n$ and its Maclaurin series for $n = 0.5$. How many terms do you need to take for the graphs to be the same between $x = 0$ and $x = 1$?

8.4   TAYLOR SERIES

The Maclaurin series for a function is valid for values of x close to zero and finding it depends on you being able to calculate the value of the function and its derivatives at zero. For a function such as $\ln(x)$ it is impossible to find a Maclaurin series because neither the function or its derivatives are defined when $x = 0$. A *Taylor series* is similar to the Maclaurin series, but is valid close to values of x other than zero. The Maclaurin series is actually a special case of the Taylor series.

The form of a Taylor series about a is

$$f(x) = a_0 + a_1(x-a) + a_2(x-a)^2 + a_3(x-a)^3 + \cdots$$

To obtain the values of the coefficients a_0, a_1, etc, we differentiate repeatedly, in the same way as for the Maclaurin series, and substitute in $x = a$ to the function and its derivatives; differentiating

$$f'(x) = a_1 + 2a_2(x-a) + 3a_3(x-a)^2 + \cdots$$
$$f''(x) = 2a_2 + 6a_3(x-a) + \cdots$$
$$f'''(x) = 6a_3 + \cdots$$

Substituting $x = a$ gives

$$a_0 = f(a)$$

$$a_1 = f'(a)$$

$$a_2 = \frac{f''(a)}{2!}$$

$$a_3 = \frac{f'''(a)}{3!}$$

$$a_n = \frac{f^n(a)}{n!} \ .$$

So the Taylor series is given by

$$f(x) = f(a) + f'(a)(x-a) + \frac{f''(a)(x-a)^2}{2!} + \frac{f'''(a)(x-a)^3}{3!} + \cdots \ .$$

Notice that if $a = 0$ this reduces to the Maclaurin series.

Example 8C

(i)   Find the first three non-zero terms of the Taylor series expansion for $\ln(x)$
        about $x = 1$.

(ii)   Use the series to find $\ln(1.1)$.

Solution

(i)   The first step is to obtain the derivatives of the function and substitute $x = 1$
        into these and the functions

$$f(x) = \ln(x) \qquad\qquad f(1) = \ln(1) = 0$$

$$f'(x) = \frac{1}{x} \qquad\qquad f'(1) = \frac{1}{1} = 1$$

$$f''(x) = -\frac{1}{x^2} \qquad\qquad f''(1) = -\frac{1}{1^2} = -1$$

$$f'''(x) = \frac{2}{x^3} \qquad\qquad f'''(1) = \frac{2}{1^3} = 2$$

The Taylor series about $x = 1$ is then given by

$$f(x) = f(1) + f'(1)(x-1) + \frac{f''(1)(x-1)^2}{2!} + \frac{f'''(1)(x-1)^3}{3!} + \cdots$$

$$= 0 + 1(x-1) + \frac{(-1)(x-1)^2}{2} + \frac{2(x-1)^3}{6} + \cdots$$

$$= (x-1) - \frac{(x-1)^2}{2} + \frac{(x-1)^3}{3} + \cdots$$

(ii)   To evaluate $\ln(1.1)$ simply substitute $x = 1.1$ into the Taylor series

$$\ln(1.1) = (1.1-1) - \frac{(1.1-1)^2}{2} + \frac{(1.1-1)^3}{3} + \cdots$$

$$= 0.1 - \frac{0.1^2}{2} + \frac{0.1^3}{3} + \cdots$$

$$= 0.1 - 0.005 + 0.0003$$

$$= 0.0953.$$

DERIVE ACTIVITY 8D

It is straightforward to obtain a Taylor or Maclaurin series in DERIVE. This process is first described and then used to illustrate some of the properties of these series.

(A) (i)   First **Author** the function for which you want to find the series. As a first example use sin$x$. Now use the **Calculus** and **Taylor** commands. You will be asked for the degree (the highest power) of the series and the point about which it is to be expanded. Initially use 5 and 0. Then **Simplify** the expression you obtain. Explain why the series you obtain is a Maclaurin series.

(ii) Now plot the original function and its Maclaurin series. What can you say about the range of validity of the approximation?

(iii) Investigate how including more terms improves the range of validity for the approximation.

(iv) What happens if you expand about $\pi/2$, π or 2π rather than 0?

(B) (i)   Find the Maclaurin series for e^x. Plot it and compare it with the original function.

(ii) Investigate how increasing the number of terms improves the quality of the approximation.

(iii) Try using a Taylor series expanded with a negative value for a. How does this affect the goodness of the approximation?

(C) Try to find a series approximation for $\ln(x)$. What problems do you encounter?

(D) Find the Taylor series for $\dfrac{1}{1+x}$ expanded about $x = -0.5$ and $x = 0.5$. Compare the graph of the function and the graph of the first four non-zero terms of each of the Taylor series. Suggest a range of values of x for which the series expansion is a good one. Are there any values of x for which a Taylor series expansion of $\dfrac{1}{1+x}$ does not exist?

Exercise 8C

1.   Find the first 3 non-zero terms of the Taylor series for

      (a) $\cos(x)$ expanded about $\dfrac{\pi}{3}$,

      (b) $\tan(x)$ expanded about $\dfrac{\pi}{4}$,

      (c) $\dfrac{1}{x}$ expanded about 1.

2.   (a)   Find the Maclaurin series for $\sin(2x)$.

      (b)   Write down the Maclaurin series for $\sin(3x)$.

3.   (a)   Explain why it is possible to find a Maclaurin series for $\ln(\cos x)$, but not
             for $\ln(\sin x)$.

      (b)   Find the first 5 non-zero terms of the series for $\ln(\cos x)$.

      (c)   Use your series to find an approximation for

$$\int_0^{0.1} \frac{\ln(\cos x)}{x^2}\,dx \ .$$

4.   Use a Maclaurin series to help you find an approximation for

$$\int_{0.5}^{1} \frac{e^x}{x}\,dx \ .$$

      What criteria did you use for your choice of degree of the Maclaurin series
      expansion of e^x?

5.   Use a Maclaurin series to help you find an approximation for

$$\int_0^1 e^{x^2}\,dx \ .$$

      What criteria did you use for your choice of degree of the Maclaurin series
      expansion of e^{x^2}?

8.5   NUMERICAL INTEGRATION

The direct and indirect (by substitution) methods of integration described in Chapter 7 provide exact, usually called *analytical*, solutions of integrals. Often in scientific and technological applications we need to evaluate the integral of a function which is given in tabular form so that the function to be integrated is not known. We need a method of integration just using these numerical values. Furthermore, it is often necessary to integrate a function for which the analytical methods do not work.

For example, $\int e^{x^2} \, dx$ does not have an analytical solution.

Considering integrals as areas under graphs allows us to develop approximate methods of integration. In the introduction to integration the areas under graphs were estimated by means of summing the areas of rectangles (see section 7.1). An alternative approach is to use trapeziums.

Consider the integral $\int_a^b f(x)dx$ and suppose that we divide the interval [a,b] in n equal subintervals each of width $h = (b-a)/n$. Approximate the curve by a set of trapeziums as shown.

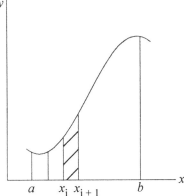

Figure 8.4

The area of the trapezium shown shaded is

$$\frac{1}{2}[f(x_i) + f(x_{i+1})](x_{i+1} - x_i) \ .$$

The total area of all the trapeziums approximating the curve is

$$s = \sum_{i=0}^{n-1} \tfrac{1}{2} h \left[f(x_i) + f(x_{i+1}) \right]$$

where $x_0 = a$ and $x_n = b$.

If we write out some of these terms we can simplify this formula.

$$s = \tfrac{1}{2} h \left\{ \left[f(x_0) + f(x_1) \right] + \left[f(x_1) + f(x_2) \right] + \cdots \cdots + \left[f(x_{n-1}) + f(x_n) \right] \right\} .$$

Notice that in this formula all the function values appear twice except $f(x_0) = f(a)$ and $f(x_n) = f(b)$. So we can write

$$s = \tfrac{1}{2} h \left\{ f(a) + 2 f(x_1) + 2 f(x_2) + \cdots \cdots 2 f(x_{n-1}) + f(b) \right\} .$$

The *trapezoidal method of integration* assumes that

$$\int_a^b f(x) dx \simeq s .$$

As the number of subintervals increases and their width decreases the approximation is improved.

Example 8D

Use the trapezoidal method with four subintervals to evaluate $\int_0^1 e^x \, dx$.

Solution

If $a = 0$ and $b = 1$ and there are four subintervals then $h = 0.25$. The formula for the trapezoidal method gives

$$\int_0^1 e^x \, dx \simeq \frac{0.25}{2} \left\{ e^0 + 2 e^{0.25} + 2 e^{0.5} + 2 e^{0.75} + e^1 \right\}$$
$$= 1.7272 \text{ to four decimal places}$$

The exact solution is $e - 1 = 1.7183$ to four decimal places so that the error is 0.52%. The trapezoidal method with four intervals is correct to two significant figures.

The trapezoidal method is just one of many methods for approximating integrals. Another method is called *Simpson's rule* and is based on approximating the function by a set of quadratic polynomials instead of straight lines.

Consider an integral $\int_a^b f(x)dx$; divide the interval $[a,b]$ into an *even* number of subintervals and define a set of quadratic polynomials such that $p(x) = f(x)$ at the points x_{i-1}, x_i and x_{i+1}.

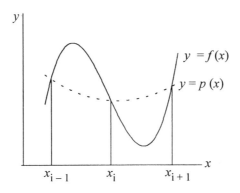

Figure 8.5

If the quadratic polynomial is $p(x) = ax^2 + bx + c$ then

$$ax_{i-1}^2 + bx_{i-1} + c = f(x_{i-1})$$
$$ax_i^2 + bx_i + c = f(x_i)$$
$$ax_{i+1}^2 + bx_{i+1} + c = f(x_{i+1})$$

Now if we integrate $p(x)$ over the subinterval $[x_{i-1}, x_{i+1}]$ and simplify the result we obtain

$$\int_{x_{i-1}}^{x_{i+1}} p(x)dx = \frac{a}{3}(x_{i+1} - x_{i-1})^3 + \frac{b}{2}(x_{i+1} - x_{i-1})^2 + c(x_{i+1} - x_{i-1}).$$

Solving for the coefficients a, b and c and substituting gives

$$\int_{x_{i-1}}^{x_{i+1}} p(x)dx = \frac{h}{3}\left(f(x_{i-1}) + 4f(x_i) + f(x_{i+1})\right).$$

(The algebra involved is somewhat tedious so the result is stated and not derived).

Adding the contributions for each pair of subintervals gives

$$\int_a^b f(x)dx \cong \frac{h}{3}\{f(x_0)+4f(x_1)+2f(x_2)+4f(x_3)+\text{LL}\quad +4f(x_{n-1})+f(x_n)\}\ .$$

where $f(x_0) = f(a)$ and $f(x_n) = f(b)$. This is called *Simpson's rule*. A convenient way of remembering this formula is

$$\int_a^b f(x)dx \cong \frac{h}{3}\ (\text{first} + \text{last} + 4 \times \text{odd values of } f + 2 \times \text{even values of } f).$$

Example 8E

Use Simpson's rule with four subintervals to evaluate $\int_0^1 e^x\ dx$.

Solution

With four subintervals we have $h = 0.25$. The four subintervals and corresponding values of the function $f = e^x$ are shown in Figure 8.6.

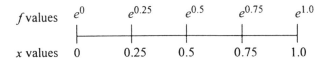

Figure 8.6

Simpson's rule with $h = 0.25$ gives

$$\int_0^1 e^x\ dx \cong \frac{0.25}{3}\left(e^0+e^1+4\left(e^{0.25}+e^{0.75}\right)+2\,e^{0.5}\right)\ .$$
$$= 1.7183\ \text{to four decimal places.}$$

Simpson's rule with four subintervals is now equal to the exact solution to four decimal places.

Exercise 8D

1. Use the trapezoidal method and Simpson's rule to approximate the following
    integrals.

    (a) $\int_{-1}^{1} e^{x^2}\, dx$ with 8 subintervals

    (b) $\int_{0}^{1} \frac{2}{1+x^2}$ with 4 subintervals

    (c) $\int_{2}^{3} \frac{1}{x}\, dx$ with 6 subintervals

2. Evaluate the integral

    $$I = \int_{0.2}^{1} \frac{\sin(x)}{x}\, dx$$

    using Simpson's rule with (a) two, (b) four, and (c) six subintervals. Compare
    your results.

3. Find an approximate value correct to three decimal places for the integral

    $$I = \int_{0}^{\pi/2} \frac{x \sin(x)}{1 + \cos^2(x)}\, dx$$

    using the trapezoidal method.

4. A function f is given in tabular form as

x	0.8	1.0	1.2	1.4	1.6	1.8	2.0
$f(x)$	4.132	5.721	6.013	7.192	8.270	9.314	10.91

    Use (a) the trapezoidal method, and (b) Simpson's rule to approximate the
    integral $\int_{0.8}^{2.0} f(x)\, dx$ with $h = 0.2$.

5.   DERIVE attempts to evaluate definite integrals by trying to find a closed form
      antiderivative and then subtract its value at the lower limit from its value at the
      upper limit. This can lead to problems.

      (a)   Evaluate $\int_{-1}^{1} \frac{1}{x^2} dx$ using **Calculus Integrate** and **Simplify**.

      (b)   **Plot** $\frac{1}{x^2}$. Does your answer to (a) make sense?

      (c)   Repeat (a) but use **approX** instead of **Simplify**. What happens? Can
             you explain the response from DERIVE?

To approximate a definite integral, DERIVE uses a form of Simpson's rule. As part
of the process DERIVE will give a warning for integrals whose values do not exist.

6.   Use DERIVE to approximate the following integrals. Which integrals do not
      exist? Can you explain why, by studying the graphs of the functions to be
      integrated?

      (a) $\int_{0.2}^{3.1} \frac{1}{x^2} dx$

      (b) $\int_{0}^{1} \frac{1}{\sqrt{x}} dx$

      (c) $\int_{-1}^{1} \frac{1}{x} dx$

      (d) $\int_{0}^{2} \frac{1}{(x-1)^2} dx$

9

Differential equations

9.1 FORMING DIFFERENTIAL EQUATIONS

Often when solving problems you may need to use information about a rate of change to form an equation.

For example, Newton's Law of Cooling states that an object will cool at a rate proportional to the difference between its temperature and the temperature of the surroundings. If T is the temperature of the object and T_0 is the temperature of the surroundings, then the temperature difference is $T - T_0$. The rate at which the temperature changes is given by $\dfrac{dT}{dt}$. As the rate is proportional to the temperature difference, we have

$$\frac{dT}{dt} \alpha (T - T_0)$$

or $\qquad \dfrac{dT}{dt} = k(T - T_0)$

where k is a constant.

This equation is known as *a first order differential equation* because it involves a first derivative. The equation below is known as *a second order differential equation* because the highest derivatives in the equation are second order.

$$\frac{d^2x}{dt^2} + \frac{dx}{dt} + 6x = \sin(t)$$

Using statements like Newton's Law of Cooling it is possible to form differential equations. In this chapter we will be mainly concerned with first order differential equations.

Example 9A

The population of a country increases at a rate that is proportional to the present population. It is increasing at a rate of 1 million per year when the population is 30 million. Use this information to form a differential equation.

Solution

Let P be the population of the country at time t. Since the rate of increase of the population, $\dfrac{dP}{dt}$, is proportional to the present population P, we have

$$\frac{dP}{dt} \propto P$$

or
$$\frac{dP}{dt} = kP$$

where k is a constant.

When $P = 30$, $\dfrac{dP}{dt} = 1$, so that

$$1 = k \times 30$$

$$k = \frac{1}{30}.$$

With this value of k the differential equation becomes

$$\frac{dP}{dt} = \frac{P}{30}.$$

Example 9B

A parachutist of mass 60kg has a terminal speed of 4 ms^{-1}. Assume that the acceleration due to gravity is 10 ms^{-2} and that the air resistance is proportional to the speed of the parachutist. Form a differential equation involving $\dfrac{dv}{dt}$ where v is the speed of the parachutist at time t.

Solution

As gravity acts down and air resistance up, the resultant force is given by $mg - R$, where R is the resistance force. Substituting the values for m and g, and noting that R is proportional to v gives

$$600 - kv$$

for some constant k.

Newton's second law states that $F = ma$, and using a as $\dfrac{dv}{dt}$ gives

$$600 - kv = 60\frac{dv}{dt} \ .$$

At the terminal speed of 4 ms^{-1}, $\dfrac{dv}{dt} = 0$ as the speed is constant. This can be used to find k

$$600 - 4k = 0$$
$$= 150 \ .$$

So the differential equations becomes

$$\frac{dv}{dt} = 10 - \frac{150v}{60}$$
$$= 10 - 2.5v \ .$$

Example 9C

The volume of a sphere is being reduced at a constant rate. Form a differential

equation for the rate of change of the radius $\dfrac{dr}{dt}$.

Solution

As the rate of decrease of the volume is constant

$$\frac{dV}{dt} = -k \ .$$

Note that because V is decreasing as t is increasing, $\dfrac{dV}{dt}$ must be negative.

Using the chain rule

$$\frac{dV}{dt} = \frac{dV}{dr} \times \frac{dr}{dt} = -k \ .$$

The volume of a sphere is given by $V = \frac{4}{3}\pi r^3$.

So $$\frac{dV}{dr} = 4\pi r^2 \ .$$

Now replacing $\dfrac{dV}{dr}$ by $4\pi r^2$ gives

$$4\pi r^2 \times \frac{dr}{dt} = -k$$

or $$\frac{dr}{dt} = -\frac{k}{4\pi r^2} \ .$$

So the radius decreases at a rate inversely proportional to the radius squared.

Exercise 9A

1.   For each of the following situations, choose a suitable variable, give the
      differential equation which it satisfies, and find the constant of proportionality.

      (a)   In an electrical circuit, the current is reducing at a rate which is
             proportional to the current flowing at that time. When the current is 40
             milliamps, it is falling at a rate of ½ milliamp per second.

      (b)   A hot object cools at a rate which is proportional to the difference
             between its temperature and that of the surrounding air. The
             surroundings remain at a constant temperature of 15 °C, and when the
             temperature of the object is 150 °C it is cooling at a rate of 12 °C per
             minute.

      (c)   A tank made of porous material has a square base of side 2 metres and
             vertical sides. The tank contains water which seeps out through the base
             and sides at a rate proportional to the total area in contact with the
             water. When the depth is 3 metres, it is observed to be falling at a rate
             of 0.2 metres per hour.

      (d)   The rate at which a substance is deposited from a solution to form a
             crystal is proportional to the product of the mass already deposited and
             the mass remaining in the solution. The solution originally contained 20
             grams of the substance. [Hint: If the amount already deposited is m
             grams, then the mass remaining in solution is $(20-m)$ grams].

2.   Air is escaping from a spherical balloon at a rate which is proportional to its
      surface area. When the radius is 20 cm, the air is escaping at 8000 cm^3 per
      second. Show that the radius is decreasing at a constant rate. Hence find the
      time taken for the radius to change from 20 cm to 15 cm.

3.   A cold can of drink is removed from a refrigerator where it had a temperature
      of 2 °C and placed in a warm room which is at a temperature of 22 °C. Using
      a suitable assumption, form a differential equation to describe how the can
      warms up.

4.   Water is pumped into a tank in the shape of an inverted cone at a rate of 0.3
      m^{3}s^{-1}. The cone has base radius 2 m and height 8 m. If h is the height of the
      water in the tank find an expression for $\dfrac{dh}{dt}$.

9.2   DIRECTION FIELDS

DERIVE allows us to draw a *direction field* for a first order differential equation. Using the differential equation it is possible to calculate the gradient of the function at any point. For example, the population model in Example 9A gives

$$\frac{dP}{dt} = \frac{P}{30}.$$

A *direction field* for $\frac{dP}{dt} = \frac{P}{30}$ is shown in figure 9.1.

The figure is drawn in the following way. In the (t,P) plane choose a set of convenient points and at each point draw a short line whose slope is the value of $\frac{dP}{dt}$ at that point. For example, at the point $(1,10)$ the slope of the short line is

$\frac{dP}{dt} = \frac{10}{30} = \frac{1}{3}$. The set of short lines is called the *direction field*, and is similar to compass needles placed in a magnetic field.

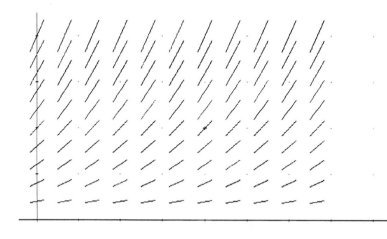

Figure 9.1   Direction field for $\frac{dP}{dt} = \frac{P}{30}$.

By looking at the direction field we can get a feel for how the population

increases with time. Notice that each vertical section of the plot is identical as $\dfrac{dP}{dt}$

depends only on P and not on t. Figure 9.2 shows the same plot, but with a number of curves superimposed on it. These curves are all possible solutions of the differential equation, each representing a different initial population. Notice how they match the gradients of the lines forming the direction fields.

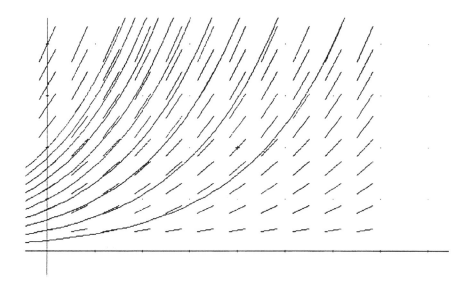

Figure 9.2   Some solution curves for $\dfrac{dP}{dt} = \dfrac{P}{30}$.

The curves shown are known as *a family of curves* and each member of this family is known as *a particular solution*.

As a second example consider the differential equation

$$\frac{di}{dt} = 1 - 2i$$

which describes how the current changes in an electrical circuit. The direction field is shown in Figure 9.3. Notice how the current will increase or decrease to the steady value of 0.5.

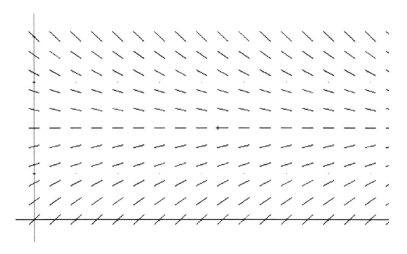

Figure 9.3   Direction field for $\dfrac{di}{dt} = 1 - 2i$

Figure 9.4 shows the same direction field with a family of curves superimposed on it. Each curve represents a different initial current.

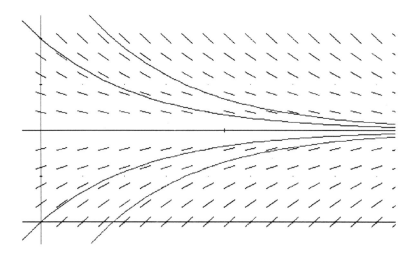

Figure 9.4   Some solution curves

DERIVE ACTIVITY 9A

This activity will show you how to draw direction fields on DERIVE and use them to obtain solutions to some problems.

(A)   (i)   **Transfer**, **Load** the **Utility** file ODE_APPR.

      (ii)   Consider the differential equation describing the current flowing in a circuit.

$$\frac{di}{dt} = 5 - 2i$$

             Author DIRECTION_FIELD (5–2i, t, 0, 5, 10, i, 0, 5, 10). This commanded is explained fully in the next section.
             Then **approX** this expression, which will take some time. The expression you obtain can be plotted to form a direction field, by following the instructions in the next section.

      (iii) Use a **Plot** window which **Overlays** the algebra window. From **Options**, **State** select **Connected** and **Small**, before plotting. **Move** the cursor to (2,2.5) and **Centre**. Also use **Scales** of 0.5 for x and 2 for y. You should now have a plot of the direction field shown in the screen dump in figure 9.5.

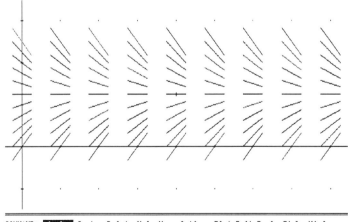

COMMAND: █Algebra█ Center Delete Help Move Options Plot Quit Scale Ticks Window
              Zoom
Enter option
Cross x:2   y:2.5   Scale x:0.5   y:2   Derive 2D-plot

Figure 9.5

(iv) What value does the current in the circuit approach as time increases?

(v)   Return to the **Algebra** window and **Author** VECTOR($2.5-2a\mathrm{e}^{-2t},a,-5,5$).
      Simplify and **Plot** this expression to produce a family of curves. The
      equation of this family of curves is

$$y = 2.5 - 2a\mathrm{e}^{-2t}$$

where a can take any value. In the plot a takes values, -5, -4, ... 5.
This equation involving the unknown constant a is called the *general
solution of the differential equation*. Each of the curves drawn that
correspond to a particular value of a are known as *particular solutions*.

(B)   (i)   A population grows so that

$$\frac{dP}{dt} = 0.05P$$

where P is the present population.

Author DIRECTION_FIELD($0.05P$, t, 0, 100, 10, P, 0, 200, 10). This
will produce a direction field where t varies from 0 to 100 in 10 steps
and P varies from 0 to 200 in 10 steps. **approX** this expression and then
Plot it using scales of 10 for x and 50 for y, with the **Centre** of the
screen at (50,100).

(ii) Return to the **Algebra** window and **Author** VECTOR($10a\mathrm{e}^{0.05t},a,0,100$).
      approX and **Plot** this expression to produce a family of curves. The
      equation of this family of curves is given by

$$P = A\mathrm{e}^{0.05t}$$

where A takes values 0, 10, 20, ..., 100. The equation above gives the
general solution while each of the curves are particular solutions.

(C)   (i)   A sample of radioactive material decays so that

$$\frac{dm}{dt} = -0.1m \ .$$

Author DIRECTION_FIELD($-0.1m$, t, 0, 20, 10, m, 0, 10, 10) then **approX** and **Plot** to produce a direction field where t varies from 0 to 20 in 10 steps and m varies from 0 to 10 in 10 steps. Choose your own **Scale** values and position for the **Centre** of the Plot.

(ii)   **Author** the expression VECTOR($ae^{-0.1t}$, a, 5, 15) to produce a family of curves that you can **Plot** over the direction field. The general solution of this differential equation is

$$m = Ae^{-0.1t}.$$

(D)   (i)   A parachutist falls so that

$$\frac{dv}{dt} = 10 - v \ .$$

Plot a direction field for $0 \le t \le 10$ and $0 \le v \le 20$. What is the terminal speed of the parachutist?

(ii)   A radioactive substance decays so that

$$\frac{dm}{dt} = -\frac{m}{2}$$

where m is the mass of the radioactive substance. Draw a direction field for $0 \le t \le 8$ and $0 \le m \le 4$. Show that $m = 4e^{-t/2}$ is a particular solution by plotting it over the direction field.

(iii)   Plot the direction field for the differential equation

$$\frac{dy}{dx} = \frac{y}{x} \ .$$

What shape do you think that the solution curves will have?

9.3   SEPARATION OF VARIABLES

Some differential equations can be quite difficult to solve, but many can be solved quite easily using a method known as *separation of variables*. If a differential equation can be written in the form

$$\frac{dy}{dx} = f(x)g(y)$$

then it can be solved by this method. For example

$$\frac{dy}{dx} = y\sin x, \text{ has } f(x) = \sin x \text{ and } g(y) = y.$$

$$\frac{dy}{dx} = 3y^2, \text{ has } f(x) = 3 \text{ and } g(y) = y^2.$$

To solve a differential equation of this type we first rearrange the equation into the form

$$\frac{1}{g(y)}\frac{dy}{dx} = f(x)$$

This process is known as *separating the variables*. Integrating each side with respect to x then gives

$$\int \frac{1}{g(y)}\frac{dy}{dx}dx = \int f(x)dx$$

$$\int \frac{1}{g(y)}dy = \int f(x)dx .$$

Integrating both sides of the equation will lead to a general solution that will contain a constant of integration. If initial values of x and y are known, then they can be used to determine the constant of integration, to give a particular solution to the differential equation.

Example 9D

The population of a country, P, is millions, grows so that

$$\frac{dP}{dt} = \frac{P}{50} .$$

(a)   Find the general solution of this differential equation.

(b)   Find a particular solution if $P = 100$ when $t = 0$.

Solution

(a)   Rearranging gives

$$\frac{1}{P}\frac{dP}{dt} = \frac{1}{50}$$

and then integrating gives

$$\int \frac{1}{P}\,dP = \int \frac{1}{50}\,dt$$

$$\ln P = \frac{t}{50} + c .$$

Notice that it is only necessary to introduce one constant of integration. Rearranging gives

$$P = e^{(t/50+c)}$$

which can be written as

$$P = Ae^{t/50}$$

where $A = e^c$.

So the *general solution* is

$$P = Ae^{t/50} .$$

(b)   To find the particular solution we substitute $P = 100$ and $t = 0$, to
       obtain

$$100 = Ae^0$$

so $A = 100$.

Using this value of A gives the *particular solution*

$$P = 100e^{t/50} .$$

Example 9E

Newton's Law of cooling leads to the differential equation

$$\frac{dT}{dt} = -5(T - 20) .$$

Solve this differential equation given that $T = 80$ when $t = 0$.

Solution

Rearranging gives

$$\frac{1}{T - 20} \frac{dT}{dt} = -5$$

and integrating each side

$$\int \frac{1}{T - 20} dT = \int -5 dt$$
$$\ln(T - 20) = -5t + c .$$

Rearranging to make T the subject of the equation gives

$$T - 20 = e^{(-5t+c)}$$

$$T = 20 + Ae^{-5t} \quad \text{(where } A = e^c\text{)}.$$

This is the general solution. The initial values can now be substituted to determine the value of A. In this case $T = 80$ when $t = 0$, giving

$$80 = 20 + Ae^0$$

so that

$$A = 60 .$$

The particular solution is

$$T = 20 + 60e^{-5t} .$$

Example 9F

The differential equation below models a body falling vertically subject to air resistance.

$$\frac{dv}{dt} = 10 - 0.2v .$$

(a)   Find the general solution of the differential equation.

(b)   If the body starts at rest find a particular solution.

Solution

(a)   Rearranging gives

$$\frac{1}{10 - 0.2v} \frac{dv}{dt} = 1$$

and integrating each side we have

$$\int \frac{1}{10 - 0.2v} dv = \int 1 dt$$

$$\frac{\ln(10 - 0.2v)}{-0.2} = t + c .$$

Rearranging to make v the subject gives

$$\ln(10 - 0.2v) = -0.2(t + c)$$
$$10 - 0.2v = e^{-0.2(t+c)} = A e^{-0.2t} \quad (\text{where } A = e^{-0.2c})$$

Hence solving for v,

$$v = \frac{10 - A e^{-0.2t}}{0.2}$$
$$= 50 - 5A e^{-0.2t}.$$

This is the general solution of the differential equation.

(b)   Since the body starts at rest, $v = 0$ when $t = 0$. Substituting these values into the general solution gives

$$0 = 50 - 5A e^{0}$$

so $A = 10$.

So the particular solution is

$$v = 50 - 50 e^{-0.2t}$$
$$= 50(1 - e^{-0.2t}).$$

Not all first order differential equations can be solved using this method. For example, in the differential equation

$$\frac{dy}{dx} = \sin(x + y)$$

the function on the right hand side $\sin(x+y)$ cannot be separated into a product $f(x)g(y)$. There are other analytical methods which are beyond the scope of this book.

DERIVE ACTIVITY 9B

Before starting this investigation set up DERIVE in word input mode, by selecting
Options Input and switching from Character to Word.

(A)   Consider the equation $\dfrac{dy}{dx} = ky$ where k is a constant.

This is an example of a simple first order differential equation for which the
variables can be separated.

$$\frac{1}{y}\frac{dy}{dx} = k \ .$$

Integrating each side gives

$$\int_{y_0}^{y} \frac{1}{y}\,dy = \int_{x_0}^{x} k\,dx \ .$$

The lower limits, x_0 and y_0, on the integral signs are known as *initial conditions*
and represent one state of the system modelled by the differential equation.

(i)   **Author** $1/y$ and use the commands **Calculus Integrate** y inserting y_0 as
the lower limit and y as the upper limit to obtain

$$\int_{y_0}^{y} \frac{1}{y}\,dy \ . \qquad \text{Do \underline{not} simplify this yet!}$$

(ii)   Now **Author** k and command **Calculus Integrate** x inserting x_0 as the
lower limit and x as the upper limit, to obtain

$$\int_{x_0}^{x} k\,dx \ .$$

(iii)   Now, by using the F3 key, **Author** the expression below

$$\int_{y_0}^{y} \frac{1}{y} dy = \int_{x_0}^{x} k \, dx \ ,$$

and **Simplify** to obtain

$$\ln(y) - \ln(y_0) = kx - kx_0 \ .$$

(iv)   Use the **soLve** command with y to obtain the general solution

$$y = y_0 \, e^{k(x-x_0)} \ .$$

(v)   Suppose that when $x = 0$, $y = 10$. This information gives the initial conditions $x_0 = 0$ and $y_0 = 10$. Use **Manage Substitute** to enter these values, and **Simplify** to obtain the solution

$$y = 10e^{kx} \ .$$

(B)   (i)   **Transfer Load** the **Utility** file ODE1.

(ii)   To solve the differential equation

$$\frac{dy}{dx} = x^2 y \ .$$

Author SEPARABLE(x^2,y, x, y, x_0, y_0), **Simplify** and **soLve** for y to obtain the general solution.

(iii)   Use the **Manage Substitute** commands to substitute $x_0 = 0$ and $y_0 = 12$. Use **Simplify** to obtain a general solution. **Plot** this solution.

(iv)   Use DERIVE to find the value of y when $x = 10$.

(v)   For what value of x is $y = 50$. (You need not consider any complex values).

Summary

Any differential equation of the form

$$\frac{dy}{dx} = f(x)g(y)$$

can be solved with $\text{SEPARABLE}(f(x), g(y), x, y, x_0, y_0)$, where x_0 and y_0 are the initial values of x and y.

(C)   Solve each of the problems below using DERIVE.

    (i)   A sky diver falls so that his speed changes according to

$$\frac{dv}{dt} = 10 - 0.3v \ .$$

    Solve this differential equation given that $v = 0$ when $t = 0$. Plot the solution and state the terminal speed of the sky diver.

    (ii)   The temperature θ of an object changes so that

$$\frac{d\theta}{dt} = -\frac{(\theta - 40)}{2} \ .$$

    If the object is initially at 80°C, find a particular solution. How long does it take for the temperature to fall to 60°C?

    (iii)   The population, P, of a colony of insects grows so that

$$\frac{dP}{dt} = \frac{P}{4} \ .$$

    Find the time taken for the population to double in size.

Exercise 9B

1.   Find the general solution of each of the differential equations.

(a) $\dfrac{dy}{dx} = y$ (b) $\dfrac{dy}{dx} = xy$ (c) $\dfrac{dy}{dx} = x^3 y$

2.   The rate of radioactive decay of a chemical is given by

$$\frac{dm}{dt} = -5m \ .$$

(a)   Find the general solution to this equation.

(b)   Given that $m = 10$ when $t = 0$ find the particular solution.

3.   Find the particular solution of each of the differential equations below.

(a) $x\dfrac{dy}{dx} = y^2$ given $y = 10$ when $x = 1$

(b) $\dfrac{dy}{dx} = x^2 y^2$ given $y = 2$ when $x = 0$

(c) $\dfrac{dy}{dx} = \dfrac{x^2}{y}$ given $y = 10$ when $x = 0$

4.   The cooling of an object is modelled by

$$\frac{d\theta}{dt} = -0.05(\theta - 20)$$

where θ is temperature in °C and t is time in minutes. If the temperature of the object is initially 80°C, find the time taken for the temperature to fall to 50°C.

5.   Solve the differential equations

(a) $\dfrac{dy}{dx} = e^{-y} \sin 2x$ given $y = 10$ when $x = 0$

(b) $\dfrac{dy}{dx} = x^2 e^y$ given $y = 1$ when $x = 0$

(c) $\dfrac{dy}{dx} = \dfrac{e^x}{y}$ given $y = 2$ when $x = 0$.

6.   Find the general solutions of these differential equations.

(a) $\dfrac{dy}{dx} = x^2(y^2 - 4)$

(b) $\dfrac{dy}{dx} = \dfrac{y+2}{x-2}$

(c) $\dfrac{dy}{dx} = x^2 y + x^2 y^2$

7.   A small particle moving in a fluid satisfies the differential equation

$$\frac{dv}{dt} = -0.2(v + v^2) \; .$$

Find the particular solution of this exercise given that $v = 40$ when $t = 0$.

8.   Which of the following differential equations can be solved by the method of separation of variables?

(a) $\dfrac{dx}{dt} = e^t x^2$ (b) $\dfrac{dP}{dx} = e^{x-P}$

(c) $\dfrac{dy}{dx} = \cos(x - y)$ (d) $\dfrac{dv}{dt} = v^2 + t^2$

9.4   NUMERICAL SOLUTIONS USING EULER'S METHOD

Euler's method provides a simple approach for finding a numerical solution to a first order differential equation of the form,

$$\frac{dy}{dx} = f(x, y) \ .$$

If we know the value of y for some value of x, say (x_0, y_0), then we can calculate $\frac{dy}{dx} = f(x_0, y_0)$ at this point. This gives the gradient of the tangent to the solution curve at (x_0, y_0) and is shown by the dotted line in figure 9.6. Further values of y can be estimated using the tangent as an approximation to the curve, and increasing x in small steps of size h.

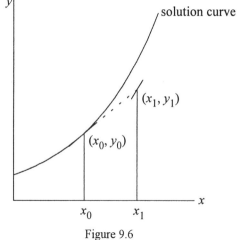

Figure 9.6

The value of y_1 is found, using

$$y_1 = y_0 + hf(x_0, y_0) \ .$$

Further steps can be taken to find approximate values of y for larger values of x. In general

$$y_{n+1} = y_n + hf(x_n, y_n) \ .$$

This formula is called *Euler's method* and is a simple iterative equation for solving first order differential equations.

Example 9G

Use Euler's method to find an approximate solution of the differential equation

$$\frac{dy}{dx} = xy$$

at the point $x = 1.5$, given the initial conditions $y = 4$ and $x = 1$.

Solution

In this case $x_0 = 1$ and $y_0 = 4$.

Using step size 0.1,

$$y_1 = y_0 + hf(x_0, y_0)$$
$$= 4 + 0.1 \times 4$$
$$= 4.4 .$$

$$y_2 = y_1 + hf(x_1, y_1) = y_1 + hx_1 y_1$$
$$= 4.4 + 0.1(1.1 \times 4.4)$$
$$= 4.884$$

Continuing this process we obtain the results shown in table 9.1.

Table 9.1

i	x_i	y_i	$f(x_i, y_i)$
0	1	4	4
1	1.1	4.4	4.84
2	1.2	4.884	5.8608
3	1.3	5.4701	7.1111
4	1.4	6.1812	8.6537
5	1.5	7.0466	

Note that the exact solution often called the *analytic solution* to this problem is
$y = A e^{0.5x^2}$ where $A = 4e^{-0.5}$ so that when $x = 1.5$

$$y = (4e^{-0.5})e^{0.5 \times 1.5^2} = 7.47 .$$

To obtain greater accuracy with Euler's method we could reduce the step size h.

DERIVE ACTIVITY 9C

(A) (i) **Transfer Load** the **Utility** file ODE_APPR.

To solve the differential equation

$$\frac{dy}{dx} = xy$$

with initial values $x_0 = 1$ and $y_0 = 4$, **Author** EULER$(xy,x,y,1,4,0.1,5)$. This will use $h = 0.1$ and give 5 steps. Using **approX** this expression leads to a set of values.

(ii) Move to a **Plot** window and from **Options State** select Connected. Now **Plot** the points that were generated by DERIVE. Adjust the **Scale** and **Centre** to give a good plot. Consecutive pairs of points are joined by straight lines to give an approximation to the true solution.

(iii) Return to the Algebra window and **Author** $4\,e^{\frac{1}{2}(x^2-1)}$ which is the solution of the differential equation. Plot this curve. Note that the difference between the curve and the approximation increases as you move away from the point $(1,4)$.

(iv) **Author** and **approX** EULER$(xy,x,y,1,4,0.05,10)$. This gives another approximation to the same differential equation, but with a greater number of steps. Plot the set of points that you obtain. How does this compare with the earlier approximation?

(B) A parachutist is falling at 20 ms^{-1} when he opens his parachute. His speed, v, then changes according to the differential equation

$$\frac{dv}{dt} = 10 - v^{\frac{3}{2}} \ .$$

Author and **approX** EULER$\left(10 - v^{\frac{3}{2}}, t, v, 0, 20, 0.1, 20\right)$. Plot the results. What is the terminal speed of the parachutist?

(C) (i) Use the $\text{EULER}\left((x^3 + y^2)/10, x, y, 0, 1, 0.4, 10\right)$ command to find a numerical solution to

$$\frac{dy}{dx} = \frac{x^3 + y^2}{10}$$

with $x_0 = 1$, $y_0 = 1$. Now **approX** and **Plot** your results.

(ii) Repeat using 20 steps of size 0.2.

(iii) Repeat again using 40 steps of size 0.1.

(iv) Continue reducing the step size and increasing the number of steps until you obtain two identical plots.

Summary

Any first order differential equation

$$\frac{dy}{dx} = f(x, y)$$

with initial values x_0, y_0 can be solved using

$$\text{EULER}(f(x,y), x, y, x_0, y_0, h, n)$$

where h is the step size and n the number of steps.

Exercise 9C

1. Solve the differential equation

    $$\frac{d\theta}{dt} = 4t - 0.05\theta$$

    given that initially $t = 0$ and $\theta = 80$, using Euler's method with step size 1 to find the value of θ when $t = 6$.

    If the analytical solution is $\theta = 80t - 1600 + 1680\,e^{-0.05t}$ calculate the percentage error in your approximate solution.

2. Use Euler's method with a step length 0.1 to find an approximate value for y when $x = 3.7$ for the differential equation

    $$\frac{dy}{dx} = \frac{x^3 y}{10}$$

    and initial conditions $x = 3$, $y = 10$.

    Use the method of separation of variables to find an exact value for y when $x = 3.7$ and calculate the percentage error of your approximation.

3. Use Euler's method with step length 0.1 to find an approximate solution to

    $$\frac{dy}{dx} = xy^2$$

    at $x = 0.3$, if initially $x = 0$ and $y = 1$. Calculate the percentage error by finding the exact solution using the method of separation of variables.

4. Find the value of y when $x = 5$ for the differential equation

    $$\frac{dy}{dx} = \frac{x+y}{y}$$

    and initial conditions $y = 2$ when $x = 0$. Use step size 1.

10

Complex numbers

10.1 THE OCCURRENCE OF COMPLEX NUMBERS

Complex numbers often arise when solving polynomial equations. For example, if we use DERIVE to solve the following equation

$$x^4 - x^3 + 2x^2 + 2x - 4 = 0$$

the DERIVE screen looks like figure 10.1.

```
         4    3     2
1:   x  - x  + 2 x  + 2 x - 4

2:   x = 1

3:   x = -1.17950

4:   x = 0.589754 - 1.74454 î

5:   x = 0.589754 + 1.74454 î
```
```
COMMAND: Author Build Calculus Declare Expand Factor Help Jump soLve Manage
         Options Plot Quit Remove Simplify Transfer moVe Window approX
Compute time: 0.1 seconds
Solve(1)                              Free:100%            Derive Algebra
```

Figure 10.1

Two of the solutions are recognisable numbers $x = 1$ and $x = -1.17950$. The other two numbers $x = 0.589754 - 1.74454i$ and $x = 0.589754 + 1.74454i$ are examples of a new set of numbers called *complex numbers*.

DERIVE ACTIVITY 10A

Use the **soLve** command on DERIVE to find the solution of the following equations. Use **Options**, **Precision** to set DERIVE in **Mixed** mode.

$$x^2 + 2 = 0$$
$$x^2 + x + 3 = 0$$
$$x^3 + 15x - 4 = 0$$
$$x^4 - 3x^2 + 2x - 4 = 0$$
$$x^6 + x^5 - x^3 - 11x^2 + 2x - 12 = 0$$

What is the relationship between the number of solutions and the degree of the polynomial?

Do you notice a pattern in the complex numbers?

Investigate your conjectures by solving some polynomial equations of your choice.

Summary

You have probably deduced that the number of solutions equals the degree of the polynomial and that the complex numbers appear to be formed in pairs that are identical except for a sign change between the first and second terms. For example when solving $x^4 - x^3 + 2x^2 + 2x - 4 = 0$ we had the pair of complex numbers

$$0.589754 + 1.74454i \quad \text{and} \quad 0.589754 - 1.74454i.$$

The meaning of i can be identified from the example

$$x^2 + 2 = 0 .$$

Rearrange this equation to give

$$x^2 = -2$$

and take square roots

$$x = \pm\sqrt{-2}$$

which DERIVE has written as $\pm 1.4142i \ (= \pm\sqrt{2}\hat{i})$.

The symbol $\hat{i}$ represents $\sqrt{-1}$. Normally in writing $\sqrt{-1}$ we use i without a hat (and engineers use the letter j). All the solutions in the DERIVE activity were of the form $z = a + bi$. If $b = 0$ the number is a familiar one called *a real number*. If $b \neq 0$ then the number is a *complex number*.

Here a and b are real numbers; a is called the *real part* of the complex number z and written as Re(z); and b is called the *imaginary part* of the complex number z and written as Im(z).

Associated with a complex number $z = a + bi$ is the complex number $z^* = a - bi$ which is called the *complex conjugate*. You will have noticed how a complex number and its conjugate appear as pairs of solutions of polynomial equations.

Example 10A

Write down the real and imaginary parts and the complex conjugate of each of the following complex numbers.

(a) $-4 + 2i$,
(b) $3 - 6i$,
(c) $1.72 - 4.21i$

Solution

Table 10.1 shows the solutions.

Table 10.1

	complex number	real part	imaginary part	complex conjugate
(a)	$-4 + 2i$	-4	2	$-4 - 2i$
(b)	$3 - 6i$	3	-6	$3 + 6i$
(c)	$1.72 - 4.21i$	1.72	-4.21	$1.72 + 4.21i$

Exercise 10A

1.   Write down the real and imaginary parts and the complex conjugate of each of
     the following complex numbers.

     (a) $2 - 3i$,
     (b) $6 + 2i$,
     (c) $1.73 - 2.19i$,
     (d) $1.7 + 4.6i$,
     (e) $-5.17 + 1.03i$,
     (f) $-4i$,
     (g) $+17i$,
     (h) $x - yi$,
     (i) $-p + qi$.

2.   Solve these equations.

     (a) $x^2 + 2x + 3 = 0$
     (b) $x^2 + 9 = 0$
     (c) $2x^2 - x + 1 = 0$
     (d) $3x^2 + 2x + 2 = 0$

3.   Show that for any complex number $z = a + bi$,

     (a) $zz^* = a^2 + b^2$
     (b) $z + z^* = 2\mathrm{Re}(z)$
     (c) $z - z^* = 2\mathrm{Im}(z)$

     where $z^* = a - bi$.

4.   Find the quadratic equation that each pair of complex numbers satisfies.

     (a) $2 - 3i, 2 + 3i$
     (b) $0.2 + 0.5i, 0.2 - 0.5i$
     (c) $3 - 5i, 3 + 5i$
     (d) $a + bi, a - bi$

10.2   GRAPHICAL REPRESENTATION OF COMPLEX NUMBERS

The complex number $z = a + bi$ can be represented on a diagram using cartesian coordinates by the point (a,b). The line from the origin to (a,b) is often drawn to help visualise the number. For example, figure 10.2 shows the complex numbers $3 + 2i$, $-1 + i$, $-2 - 3i$ and $4 - 2i$.

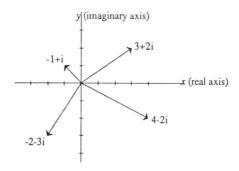

Figure 10.2   The Argand Diagram

The representation of a complex number in this way is called an *Argand Diagram*.

From the Argand diagram we can associate the size of a complex number with the length of the line and the inclination of a complex number with the angle the line makes with the x-axis. The size and inclination of a complex number are called the *modulus* and *argument* respectively.

For the complex number $z = a + bi$

modulus $= |z| = r = \sqrt{a^2 + b^2}$

argument $= \arg(z) = \theta$

       $=$ angle of OP to positive x direction.

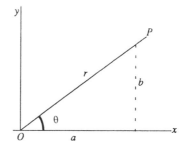

Figure 10.3

If z lies above the x-axis $0 \le \arg(z) \le \pi$ whereas if z lies below the x-axis $-\pi < \arg(z) < 0$.

Example 10B

Show the following complex numbers on an Argand diagram and find the modulus and argument of each one.

$$3 + 2i, \quad -4 - 2i, \quad 1 - 2i$$

Solution

Figure 10.4 shows the complex numbers on an Argand diagram.

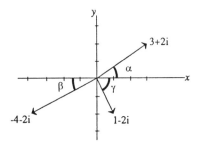

Figure 10.4

For 3 + 2i: $|3+2i| = \sqrt{3^2 + 2^2} = \sqrt{13}$

$$\arg(3+2i) = \alpha = \arctan\frac{2}{3} = 0.522$$

For −4 − 2i: $|-4-2i| = \sqrt{4^2 + 2^2} = \sqrt{20}$

$$\arg(-4-2i) = -(\pi - \beta) = -\left(\pi - \arctan\left(\frac{2}{4}\right)\right) = -2.68$$

For 1 − 2i: $|1-2i| = \sqrt{1^2 + 2^2} = \sqrt{5}$

$$\arg(1-2i) = -\gamma = -\arctan\left(\frac{2}{1}\right) = -1.11$$

Figure 10.3 shows that the relationships between a, b and r, θ are

$$a = r\cos\theta \text{ and } b = r\sin\theta.$$

Hence the complex number $z = a + bi$ can be written in terms of r and θ as

$$z = r\cos\theta + ir\sin\theta = r(\cos\theta + i\sin\theta).$$

This is called *the polar form* of the complex number.

For example for the complex number $1 - 2i$ of Example 10B the polar form is written as $\sqrt{5}(\cos 1.11 - i\sin 1.11)$ since $\cos(-1.11) = \cos(1.11)$ and $\sin(-1.11) = -\sin(1.11)$.

Exercise 10B

1.   Determine the modulus and argument of the following complex numbers. Show each of them on an Argand diagram. Write each of them in polar form.

(a) i (b) $1 + i$ (c) $2 - 3i$

(d) $-1 - 4i$ (e) $-3 - 4i$ (f) $-5 + 12i$

(g) 1 (h) $-3i$ (i) -4

(j) $1 - \sqrt{3}i$ (k) $-1 - \sqrt{3}i$ (l) $1.32 - 6.21i$

   Check your answers using DERIVE. For the modulus of z DERIVE uses ABS(z) and for the argument DERIVE uses PHASE(z). Remember to use alt-i for $\hat{\imath}$.

2.   Find the solutions of the following equations and write them in polar form.

(a) $x^2 + 2x + 2 = 0$

(b) $x^2 + 8 = 0$

3.   Use DERIVE to evaluate the following.

(a) $|3 + 4i|$ (b) $\arg(3 - 4i)$ (c) $|(1 + i)(3 - 2i)|$

(d) $\left|\dfrac{5 - 12i}{1 + i}\right|$ (e) $\arg\left(\dfrac{5 - 12i}{1 + i}\right)$ (f) $|(3 - i)^2 (2 + i)^3|$

10.3   THE ARITHMETIC OF COMPLEX NUMBERS

The addition, subtraction and multiplication of complex numbers is straightforward. For example,

$$(2-3i) + (1+4i) = 2 - 3i + 1 + 4i = (2+1) + (-3i+4i) = 3 + i$$

$$(2-3i) - (1+4i) = 2 - 3i - 1 - 4i = (2-1) + (-3i-4i) = 1 - 7i$$

$$(2-3i)(1+4i) = 2 + 8i - 3i - 12i^2 = 2 + 8i - 3i + 12 = 14 + 5i$$

The division of complex numbers uses the complex conjugate.

Consider $z = \dfrac{(2-3i)}{(1+4i)}$. We multiply top and bottom by $(1-4i)$ giving

$$z = \frac{(2-3i)(1-4i)}{(1+4i)(1-4i)} = \frac{2-8i-3i+12i^2}{1-4i+4i-16i^2} = \frac{2-8i-3i-12}{1-4i+4i+16} = \frac{-10-11i}{17}.$$

Exercise 10C

1.   For each pair of complex numbers find

$$z_1 + z_2, \quad z_1 - z_2, \quad z_1 z_2, \quad \frac{z_1}{z_2}, \quad z_1^2 \quad \text{and} \quad 3z_1 - 2z_2.$$

(a) $z_1 = i, z_2 = 3 + 4i$ (b) $z_1 = -3 + 4i, z_2 = 1 - i$

(c) $z_1 = 1 + i, z_2 = 1 - i$ (d) $z_1 = -6 - i, z_2 = 1 - 2i$

(e) $z_1 = 3 + 2i, z_2 = 5 - i$ (f) $z_1 = -3 - 4i, z_2 = 5 + 12i$

Check your answers using DERIVE.

2.   With $z = 3 - 2i$ find

(a) iz (b) $z*$ (c) $\dfrac{1}{z}$ (d) $\dfrac{1}{z*}$.

3.   Show on an Argand diagram that the effect of multiplying a complex number by i is a rotation of $\dfrac{\pi}{2}$ (ie. 90°) anticlockwise.

4.   Find the values of a and b if

(a) $a + bi = \dfrac{3}{\cos(\theta) + i\sin(\theta)}$,

(b) $a + bi = \dfrac{2 - i}{1 + \cos(\theta) - i\sin(\theta)}$.

5.   Find the real and imaginary parts of z when

$$\frac{1}{z} = \frac{1}{1+i} - \frac{1}{2-i} .$$

6.   In DERIVE use the PHASE command to:

(a)   Show that the argument of $(1+i)^3$ is three times the argument of $(1+i)$;

(b)   Show that the argument of $\left(-\dfrac{1}{2} + \dfrac{\sqrt{3}}{2} i \right)^4$ is four times the argument of

$\left(-\dfrac{1}{2} + \dfrac{\sqrt{3}}{2} i \right)$;

Propose a rule for the modulus and argument of z^n in terms of n and the modulus and argument of z. Use DERIVE to test your rule for i^6, $(2+3i)^3$, $(4-2i)^5$, $(1-i)^7$, $(4.7+1.7i)^9$.

Hence write z^n in polar form given that $z = r(\cos\theta + i\sin\theta)$.

7.   (a)   Use DERIVE to simplify the expression e^{2+3i}.

(b)   Find the modulus and arguments of the answer to part (a).

Repeat parts (a) and (b) for the expressions e^{1-i}, e^{3+4i} and $e^{0.4i}$.

What can you deduce about $\arg(e^{a+bi})$?

10.4   EULER'S FORMULA

We now bring together the results of problems 6 and 7 of Exercise 10C.

In problem 6 you may have deduced that if

$$z = r(\cos\theta + i\sin\theta)$$

then z^n can be written as

$$z^n = r^n(\cos(n\theta) + i\sin(n\theta)) \, .$$

This says that the argument of z^n is n times the argument of z. This important result is known as *de Moivre's Theorem*.

In problem 7 you will have seen that e^{a+bi} is a complex number with argument b. So we can use the polar form to write

$$e^{a+bi} = e^a e^{bi} \, .$$

Here e^a is a real number so we can interpret e^a *as* the modulus of e^{a+bi} hence we have

$$\boxed{e^{bi} = \cos b + i\sin b.}$$

This is known as *Euler's formula*. It provides a fascinating and powerful link between the exponential and trigonometric functions.

We can now write a complex number in three different ways

$$z = x + iy \qquad\qquad \textit{cartesian form}$$

$$z = r(\cos\theta + i\sin\theta) \qquad \textit{polar form}$$

$$z = re^{i\theta} \qquad\qquad \textit{exponential form.}$$

Example 10C

Use Euler's formula to prove de Moivre's Theorem.

Solution

Let $z = r(\cos\theta + i\sin\theta)$. Then using Euler's formula we have

$$z = re^{i\theta} .$$

Using the rules of indices, we have, for any n

$$z^n = (re^{i\theta})^n = r^n e^{i(n\theta)} .$$

Using Euler's formula again

$$z^n = r^n(\cos n\theta + i\sin n\theta) .$$

Example 10D

Write $-1 + i$ in the polar form and hence evaluate $(-1+i)^8$.

Solution

$$|-1+i| = \sqrt{2} \quad\text{and}\quad \arg(-1+i) = \frac{3\pi}{4}$$

so that

$$-1+i = \sqrt{2}\left[\cos(3\pi/4) + i\sin(3\pi/4)\right] .$$

Now

$$(-1+i)^8 = (\sqrt{2})^8\left[\cos(3\pi/4) + i\sin(3\pi/4)\right]^8$$

$$= 2^4\left[\cos\left(8 \times \frac{3\pi}{4}\right) + i\sin\left(8 \times \frac{3\pi}{4}\right)\right]$$

$$= 2^4\left[\cos(6\pi) + i\sin(6\pi)\right] = 2^4[1+0i]$$

$$= 16.$$

Exercise 10D

1. Write the following complex numbers in polar and exponential form.

   (a) $3 + 4i$ (b) $1 - i$ (c) $-\dfrac{\sqrt{3}}{2} - \dfrac{1}{2}i$

   (d) $5 - 12i$ (e) $-2 + 2i$ (f) $-0.3 - 0.7i$

   (g) $\dfrac{1}{3 + 4i}$ (h) $\dfrac{1}{1 + 2i}$ (i) $\dfrac{1 - i}{5 - 12i}$

2. Write the following complex numbers in cartesian form.

   (a) $e^{i\pi}$ (b) $3e^{-i\pi/2}$ (c) $e^{-i\pi}$

   (d) $2e^{0.7i}$ (e) $4e^{-0.1i}$ (f) $0.9e^{i\pi/4}$

   Show each complex number on an Argand diagram.

3. Given $z_1 = 2e^{i\pi/4}$ and $z_2 = 3e^{-i\pi/2}$ find the modulus and arguments of

   (a) z_1^2 (b) z_2^3 (c) $z_1^2 z_2^3$ (d) $\dfrac{z_1^2}{z_1^3}$.

4. Repeat problem 3 for $z_1 = 4e^{0.7i}$ and $z_2 = 0.5e^{-0.1i}$.

5. Write $3 - 4i$ in polar form and hence evaluate $(3-4i)^6$.

6. Write $1 + i$ in polar form and hence evaluate $(1+i)^{12}$.

7. By expanding $(\cos\theta + i\sin\theta)^3$ and using de Moivre's theorem, write $\cos 3\theta$ and $\sin 3\theta$ in terms of $\cos\theta$ and $\sin\theta$.

10.5 FINDING ROOTS OF COMPLEX NUMBERS

One of the most common uses of de Moivre's theorem is to find the roots of a complex number such as $\sqrt{z}$ and $\sqrt[4]{z}$. The following examples show the method of approach.

Example 10E

Find the three cube roots of $1 + i$.

Solution

The modulus of $1 + i$ is $\sqrt{2}$ and the argument of $1 + i$ is $\dfrac{\pi}{4}$, so that

$$1+i = \sqrt{2}\left[\cos\left(\frac{\pi}{4}\right)+i\sin\left(\frac{\pi}{4}\right)\right].$$

Take the cube root of each side

$$(1+i)^{\frac{1}{3}} = (\sqrt{2})^{\frac{1}{3}}\left[\cos\left(\frac{\pi}{4}\right)+i\sin\left(\frac{\pi}{4}\right)\right]^{\frac{1}{3}}$$

$$= 2^{\frac{1}{6}}\left[\cos\left(\frac{\pi}{12}\right)+i\sin\left(\frac{\pi}{12}\right)\right]$$

by de Moivre's theorem. In cartesian form one cube root of $(1+i)$ is $1.084 + 0.291i$.

To find the other two cube roots we need to observe that the argument of a complex number is not unique. Because of the periodicity of $\cos\theta$ and $\sin\theta$ we could write

$$z = r(\cos\theta + i\sin\theta) = r\left[\cos(\theta + 2\pi k) + i\sin(\theta + 2\pi k)\right]$$

where $k = 0, 1, -1, 2, -2, 3, -3, \ldots$

Figure 10.5 shows the values of $k = 1$ and $k = 2$.

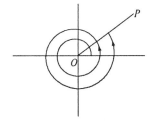

Figure 10.5

Each value of k is just a rotation of the line OP through an angle 2π. So in general $(1+i)$ can be written as

$$1+i = \sqrt{2}\left[\cos\left(\frac{\pi}{4}+2\pi k\right)+i\sin\left(\frac{\pi}{4}+2\pi k\right)\right].$$

Take cube roots of each side and apply de Moivre's theorem,

$$(1+i)^{\frac{1}{3}} = 2^{\frac{1}{6}}\left[\cos\left(\frac{\pi}{12}+\frac{2\pi k}{3}\right)+i\sin\left(\frac{\pi}{12}+\frac{2\pi k}{3}\right)\right].$$

Substituting in some values for k and simplifying we obtain

$$k=0 \quad (1+i)^{\frac{1}{3}} = 1.084+0.291i$$
$$k=1 \quad (1+i)^{\frac{1}{3}} = -0.794+0.794i$$
$$k=2 \quad (1+i)^{\frac{1}{3}} = -0.291-1.084i$$
$$k=3 \quad (1+i)^{\frac{1}{3}} = 1.084+0.291i$$
$$k=4 \quad (1+i)^{\frac{1}{3}} = -0.794+0.794i$$
$$k=5 \quad (1+i)^{\frac{1}{3}} = -0.291-1.084i$$

The formula for $(1+i)^{\frac{1}{3}}$ gives exactly three different roots corresponding to $k = 0, 1, 2$. When $k = 3, 4, 5$ or $k = 6, 7, 8$ etc the roots repeat themselves. The three roots are shown in the Argand diagram, figure 10.6.

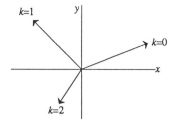

Figure 10.6   The three cube roots of $1 + i$

In general, to find the n n^{th} roots of a complex number z we write z in polar form

$$z = r\left[\cos(\theta + 2\pi k) + i\sin(\theta + 2\pi k)\right] .$$

Then

$$z^{\frac{1}{n}} = r^{\frac{1}{n}}\left[\cos\left(\frac{\theta}{n} + \frac{2\pi k}{n}\right) + i\sin\left(\frac{\theta}{n} + \frac{2\pi k}{n}\right)\right]$$

choosing $k = 0, 1, 2, 3, \ldots (n-1)$.

Example 10F

Find the four fourth roots of $3 + 4i$.

Solution

The modulus of $(3+4i)$ is 5 and the argument of $(3+4i)$ is 0.9273, so that

$$3 + 4i = 5[\cos(0.9273 + 2\pi k) + i\sin(0.9273 + 2\pi k)] .$$

Take the fourth roots of each side

$$(3+4i)^{\frac{1}{4}} = 5^{\frac{1}{4}}\left[\cos\left(\frac{0.9273}{4} + \frac{2\pi k}{4}\right) + i\sin\left(\frac{0.9273}{4} + \frac{2\pi k}{4}\right)\right] .$$

For $k = 0, 1, 2$ and 3 we have the four fourth roots as

$$
\begin{aligned}
k &= 0 & z_1 &= 1.4553 + 0.3436i \\
k &= 1 & z_2 &= -0.3436 + 1.4553i \\
k &= 2 & z_3 &= -1.4553 - 0.3436i \\
k &= 3 & z_4 &= 0.3436 - 1.4553i
\end{aligned}
$$

Exercise 10E

1.   Find the square roots and cube roots of the following complex numbers, giving
      your answers in cartesian form.

      (a) $1 - i$ (b) $2 + i$ (c) $3i$

      (d)   4   (e) $-1 + \sqrt{3}i$ (f) $-5 - 12i$

2.   Find the following complex numbers in polar form.

      (a) $(\sqrt{3} - i)^{\frac{1}{5}}$ (b) $(1 + i)^{\frac{4}{3}}$

      (c) $(-3 + 4i)^{-\frac{2}{3}}$ (d) $(-1)^{\frac{1}{6}}$

3.   If you use DERIVE to evaluate $(2 - i)^{\frac{1}{3}}$ you get the complex number
      $1.29207 - 0.201294i$. Use an Argand diagram to deduce the other two cube
      roots.

4.   Use DERIVE and an Argand diagram to find the following complex numbers.

      (a) $(3i)^{\frac{1}{4}}$ (b) $(\sqrt{3} - i)^{\frac{1}{6}}$

11

Matrices

In this chapter we provide a short introduction to the algebra of matrices and their use in solving simultaneous linear equations.

11.1 INTRODUCTION

In mathematics we often represent quantities as arrays of numbers (or symbols). For example, the pair of numbers (2,3) could represent a point in the x-y plane of a cartesian coordinate system. In chapters 1 and 2 we applied the FIT function in DERIVE to fit a curve to a set of data points displayed in the form

$$\begin{bmatrix} 1 & 7 \\ 2 & 11 \\ 3 & 15 \\ 4 & 19 \end{bmatrix}$$. These arrays are examples of *matrices.*

A *matrix* is an array of numbers arranged in regular rows and columns. Here are some examples of matrices.

$$A = \begin{bmatrix} 1 & -3 \\ 0 & 4 \\ 5 & 1 \end{bmatrix} \qquad \text{is a matrix with 3 rows and 2 columns}$$

$$B = \begin{bmatrix} -1 & 0 \\ 6 & 3 \end{bmatrix} \qquad \text{is a matrix with 2 rows and 2 columns}$$

$$C = \begin{bmatrix} a_1 & a_2 & a_3 \\ b_1 & b_2 & b_3 \end{bmatrix} \quad \text{is a matrix with 2 rows and 3 columns}$$

$$D = \begin{bmatrix} 1 & 0 & 4 & -5 & 6 \end{bmatrix} \quad \text{is a matrix with 1 row and 5 columns.}$$

In general, a matrix A with m rows and n columns is called an $m \times n$ *matrix*. A $1 \times n$ matrix is called a *row vector*; matrix D above is an example of a row vector. An

$m \times 1$ matrix is called a *column vector*; $\begin{bmatrix} 1 \\ 0 \\ 3 \end{bmatrix}$ is an example of a column vector. Often

these are just called *vectors*.

If the number of rows is equal to the number of columns then the matrix is called a *square matrix*. Matrix B above is an example of a square matrix.

A *diagonal matrix* is a square matrix with zeros everywhere except down the

leading diagonal. For example, $\begin{bmatrix} 1 & 0 & 0 \\ 0 & 4 & 0 \\ 0 & 0 & -3 \end{bmatrix}$ is a 3×3 diagonal matrix. If the

diagonal elements are all equal to 1 the matrix is called *a unit matrix*. The 4×4 unit matrix is

$$\begin{bmatrix} 1 & 0 & 0 & 0 \\ 0 & 1 & 0 & 0 \\ 0 & 0 & 1 & 0 \\ 0 & 0 & 0 & 1 \end{bmatrix}.$$

Unit matrices are denoted by the letter I.

Matrices are often used to represent a set of simultaneous linear equations such as

$$\begin{array}{rcrcrcr} 2x_1 & - & 4x_2 & + & x_3 & = & 5 \\ x_1 & + & x_2 & - & 3x_3 & = & -1 \\ 3x_1 & - & x_2 & + & x_3 & = & 0 \end{array}.$$

The properties of the matrix of coefficients $\begin{bmatrix} 2 & -4 & 1 \\ 1 & 1 & -3 \\ 3 & -1 & 1 \end{bmatrix}$ turn out to be very important

in solving the equations. The unknown values of x can be written as a column vector

$x = \begin{bmatrix} x_1 \\ x_2 \\ x_3 \end{bmatrix}$. The three linear equations can be written as

$$\begin{bmatrix} 2 & -4 & 1 \\ 1 & 1 & -3 \\ 3 & -1 & 1 \end{bmatrix} \begin{bmatrix} x_1 \\ x_2 \\ x_3 \end{bmatrix} = \begin{bmatrix} 5 \\ -1 \\ 0 \end{bmatrix}.$$

If we denote the matrix of coefficients by A and the column vector on the right hand side by b then the set of linear equations can be written as

$$Ax = b \ .$$

In general, to solve a set of simultaneous linear equations we need to solve matrix equations of the form $Ax = b$ where A is a square matrix and x and b are column vectors.

Exercise 11A

Write each of the following sets of simultaneous linear equations in matrix form.

(a) $x - y = 4$
       $2x + 3y = 1$

(b) $-2x + y = -6$
       $x - y = 4$

(c) $3x - 6y + z = 7$
       $-2x + y - 3z = 2$
       $x + y + z = 0$

(d) $4x_1 - x_2 + x_3 = 1$
       $3x_1 + x_2 - 2x_3 = -3$
       $-x_1 - 4x_2 + x_3 = 5$

(e) $x_1 - 3x_2 + x_3 + 4x_4 - x_5 = 0$
       $-x_1 - 4.1x_3 + 2x_5 = 1.3$
       $0.3x_1 - 0.7x_2 + 4.1x_4 - x_5 = 2.7$
       $1.4x_1 - x_2 + 3.1x_4 = 0.4$
       $3x_1 + x_2 - x_3 + 2x_4 - x_5 = -3.5$

11.2   THE ARITHMETIC OF MATRICES

Matrices can be added, subtracted and multiplied together but you *cannot* divide by a matrix. Matrix division has no meaning.

Equality of matrices

Two matrices A and B are *equal* if they are the same size and the

corresponding elements are equal. For example, if $A = \begin{bmatrix} a_1 & a_2 & a_3 \\ b_1 & b_2 & b_3 \end{bmatrix}$ and

$B = \begin{bmatrix} 0 & -1 & 2 \\ 4 & 1 & -1 \end{bmatrix}$ then $A = B$ means that

$$
\begin{array}{lll}
a_1 = 0 & a_2 = -1 & a_3 = 2 \\
b_1 = 4 & b_2 = 1 & b_3 = -1
\end{array}
$$

Addition and Subtraction of matrices

To add (or subtract) two matrices together they must be the same size. We then add (or subtract) corresponding elements. For example if

$$
A = \begin{bmatrix} 0 & -1 & 2 \\ 4 & 1 & -1 \end{bmatrix} \quad \text{and} \quad B = \begin{bmatrix} 3 & 2 & -1 \\ 0 & -1 & -2 \end{bmatrix}
$$

then we can add (or subtract) A and B because they are each 2×3 matrices to give

$$
A + B = \begin{bmatrix} 0 & -1 & 2 \\ 4 & 1 & -1 \end{bmatrix} + \begin{bmatrix} 3 & 2 & -1 \\ 0 & -1 & -2 \end{bmatrix}
$$

$$
= \begin{bmatrix} 0+3 & -1+2 & 2-1 \\ 4+0 & 1-1 & -1-2 \end{bmatrix} = \begin{bmatrix} 3 & 1 & 1 \\ 4 & 0 & -3 \end{bmatrix}
$$

$$A - B = \begin{bmatrix} 0 & -1 & 2 \\ 4 & 1 & -1 \end{bmatrix} - \begin{bmatrix} 3 & 2 & -1 \\ 0 & -1 & -2 \end{bmatrix}$$

$$= \begin{bmatrix} 0-3 & -1-2 & 2-(-1) \\ 4-0 & 1-(-1) & -1-(-2) \end{bmatrix} = \begin{bmatrix} -3 & -3 & 3 \\ 4 & 2 & 1 \end{bmatrix}$$

For matrix addition $A + B = B + A$. So we say that matrix addition is *commutative*. Note that $A - B \neq B - A$. So we say matrix subtraction is not commutative.

Multiplication by a single number

To multiply a matrix by a single number we multiply each element by the number. For example, for A defined above

$$-2A = -2 \begin{bmatrix} 0 & -1 & 2 \\ 4 & 1 & -1 \end{bmatrix} = \begin{bmatrix} -2 \times 0 & -2 \times -1 & -2 \times 2 \\ -2 \times 4 & -2 \times 1 & -2 \times -1 \end{bmatrix}$$

$$= \begin{bmatrix} 0 & 2 & -4 \\ -8 & -2 & 2 \end{bmatrix}$$

Example 11A

Given the matrices $B = \begin{bmatrix} 1 & 0 & 1 \\ -2 & 1 & 3 \end{bmatrix}$ and $C = \begin{bmatrix} 0 & -1 & 5 \\ 4 & 0 & 1 \end{bmatrix}$, find the matrix A if

$$3A + 2B = C.$$

Solution

Applying the normal rules of algebra we subtract $2B$ from each side to give

$$3A = C - 2B$$

and dividing by 3

$$A = \tfrac{1}{3}(C - 2B) .$$

Substituting for B and C

$$A = \frac{1}{3}\left(\begin{bmatrix} 0 & -1 & 5 \\ 4 & 0 & 1 \end{bmatrix} - 2 \begin{bmatrix} 1 & 0 & 1 \\ -2 & 1 & 3 \end{bmatrix} \right)$$

$$= \frac{1}{3} \begin{bmatrix} -2 & -1 & 3 \\ 8 & -2 & -5 \end{bmatrix}$$

$$= \begin{bmatrix} -\frac{2}{3} & -\frac{1}{3} & 1 \\ \frac{8}{3} & -\frac{2}{3} & -\frac{5}{3} \end{bmatrix} .$$

Multiplication of matrices

Two matrices A and B can be multiplied together to give a matrix AB provided the number of columns of A equals the number of rows of B. If this condition holds then AB is calculated by taking the products of every row of A with every column of B in a special way.

Example 11B

Let $A = \begin{bmatrix} 2 & -1 \\ 1 & -3 \end{bmatrix}$, $B = \begin{bmatrix} 4 & 1 & 5 \\ -3 & 0 & 7 \end{bmatrix}$ and $C = \begin{bmatrix} 1 & 0 \\ -2 & 6 \\ 5 & -1 \end{bmatrix}$. Form all the possible matrix products of A, B and C.

Solution

We could attempt to form the products

$$AB, \; BA, \; AC, \; CA, \; BC \text{ and } CB.$$

However, using the rule that the number of columns of the first matrix must equal the number of rows of the second matrix, the possible matrix products are

$$AB, \; CA \text{ and } BC.$$

To calculate AB:

$$AB = \begin{bmatrix} 2 & -1 \\ 1 & -3 \end{bmatrix} \begin{bmatrix} 4 & 1 & 5 \\ -3 & 0 & 7 \end{bmatrix}$$

$$= \begin{bmatrix} (2 \times 4 + -1 \times -3) & (2 \times 1 + -1 \times 0) & (2 \times 5 + -1 \times 7) \\ (1 \times 4 + -3 \times -3) & (1 \times 1 + -3 \times 0) & (1 \times 5 + -3 \times 7) \end{bmatrix} = \begin{bmatrix} 11 & 2 & 3 \\ 13 & 1 & -16 \end{bmatrix}$$

The method is that each number in the rth row of A is multiplied by each corresponding term in the sth column of B and added to give the element in the rth row, sth column of AB. For example, the element in the 2nd row and 3rd column of AB is shown by the following.

$$\begin{bmatrix} 2 & -1 \\ -1 & -3 \end{bmatrix} \begin{bmatrix} 4 & 1 & 5 \\ -3 & 0 & 7 \end{bmatrix} = \begin{bmatrix} * & * & * \\ * & * & c \end{bmatrix}$$

$$c = (-1 \times 5) + (-3 \times 7)$$

Similarly

$$CA = \begin{bmatrix} 1 & 0 \\ -2 & 6 \\ 5 & -1 \end{bmatrix} \begin{bmatrix} 2 & -1 \\ 1 & -3 \end{bmatrix}$$

$$= \begin{bmatrix} (1 \times 2 + 0 \times -1) & (1 \times -1 + 0 \times -3) \\ (-2 \times 2 + 6 \times 1) & (-2 \times -1 + 6 \times -3) \\ (5 \times 2 + -1 \times 1) & (5 \times -1 + -1 \times -3) \end{bmatrix} = \begin{bmatrix} 2 & -1 \\ 2 & -16 \\ 9 & -2 \end{bmatrix}$$

$$BC = \begin{bmatrix} 4 & 1 & 5 \\ -3 & 0 & 7 \end{bmatrix} \begin{bmatrix} 1 & 0 \\ -2 & 6 \\ 5 & -1 \end{bmatrix}$$

$$= \begin{bmatrix} (4 \times 1 + 1 \times -2 + 5 \times 5) & (4 \times 0 + 1 \times 6 + 5 \times -1) \\ (-3 \times 1 + 0 \times -2 + 7 \times 5) & (-3 \times 0 + 0 \times 6 + 7 \times -1) \end{bmatrix} = \begin{bmatrix} 27 & 1 \\ 32 & -7 \end{bmatrix}$$

The examples of matrix multiplication in Example 11B show several important properties:

1.   In general $AB \neq BA$. So we say that matrix multiplication is not commutative. For example,

$$\begin{bmatrix} 1 & 0 \\ 1 & 3 \end{bmatrix} \begin{bmatrix} 1 & 4 \\ -1 & 0 \end{bmatrix} = \begin{bmatrix} 1 & 4 \\ -2 & 4 \end{bmatrix}$$

$$\begin{bmatrix} 1 & 4 \\ -1 & 0 \end{bmatrix} \begin{bmatrix} 1 & 0 \\ 1 & 3 \end{bmatrix} = \begin{bmatrix} 5 & 12 \\ -1 & 0 \end{bmatrix}$$

2.   If A is an $m \times r$ matrix and B is an $r \times n$ matrix then AB is an $m \times n$ matrix.

$$A \qquad B \qquad = \qquad AB$$

$$m \times r \qquad r \times n \qquad\qquad m \times n$$

Exercise 11B

For the problems in this Exercise use the matrices given by

$$A = \begin{bmatrix} 0 & 1 \\ 1 & 0 \end{bmatrix} \qquad B = \begin{bmatrix} 1 & 3 & 0 \\ -1 & 2 & 1 \end{bmatrix} \qquad C = \begin{bmatrix} 1 & 4 & 2 \\ 0 & -2 & 1 \\ -1 & 1 & 0 \end{bmatrix}$$

$$D = \begin{bmatrix} 0 & 1 \\ 1 & 2 \\ -1 & -3 \end{bmatrix} \qquad E = \begin{bmatrix} 0 & 1 & 3 \\ -1 & 0 & 2 \\ 0 & 1 & 2 \end{bmatrix} \qquad F = \begin{bmatrix} 0 & 1 & -1 & 3 \\ -1 & 2 & 3 & 0 \\ 4 & 0 & -1 & 2 \end{bmatrix}$$

$$G = \begin{bmatrix} 1 & 0 \\ 0 & 1 \end{bmatrix}$$

and I is the unit matrix of appropriate size.

1.   Calculate each of the following (where possible).

   (a) $C + E$ (b) $A + B$ (c) $-3E$

   (d) $A + I$ (e) $A - I$ (f) $2C - 3E$

   (g) $E + F$ (h) AB (i) AC

   (j) DE (k) ED (l) EF

   (m) FC (n) BC (o) AI

   (p) BI (q) IB (r) DI

2.   If x is the column vector $\begin{bmatrix} x_1 \\ x_2 \\ x_3 \end{bmatrix}$ then expand Bx, Cx and Ex.

3.   Find the matrix X if

   (a) $X + C = E$

   (b) $X + 3C = 5E$

   (c) $2X - C = 2E$

4.   If $\begin{bmatrix} a & 1 \\ 1 & b \end{bmatrix} A = I$, find a and b.

5.   Show that if X is an $m \times m$ square matrix and if I is an $m \times m$ unit matrix then

   $$XI = IX = X.$$

11.3 DERIVE AND MATRICES

This section introduces the matrix algebra commands of DERIVE. Using DERIVE for the arithmetic of matrices saves time in the same way as we use a calculator to manipulate numbers.

DERIVE ACTIVITY 11A

(A)   To enter a matrix you use the **Declare** command.

Consider the matrix $A = \begin{bmatrix} 1 & 0 & 3 \\ -1 & 2 & 4 \end{bmatrix}$.

Declare Matrix gives a request for the number of rows and columns.
Enter 2 for the number of rows and 3 for the number of columns.

DERIVE now requests the entries of the matrix. Insert the following sequence of numbers pressing enter after each one: 1 0 3 −1 2 4.

You should now see the matrix A as line #1.

You can label this matrix as a by using the **Declare Function** commands. Respond to

DECLARE FUNCTION name: _ with A
DECLARE FUNCTION value: _ with #1

You should now see the DERIVE screen shown in figure 11.1.

1: $\begin{bmatrix} 1 & 0 & 3 \\ -1 & 2 & 4 \end{bmatrix}$

2: a := $\begin{bmatrix} 1 & 0 & 3 \\ -1 & 2 & 4 \end{bmatrix}$

```
COMMAND: Author Build Calculus Declare Expand Factor Help Jump soLve Manage
         Options Plot Quit Remove Simplify Transfer moVe Window approX
Enter option
a():=                              Free:100%              Derive Algebra
```

Figure 11.1

(B)   Now enter the following matrices as lines #3 and #5

$$B = \begin{bmatrix} 0 & -2 & 1 \\ -1 & 1 & 3 \end{bmatrix} \qquad C = \begin{bmatrix} 0 & -1 \\ 2 & 3 \\ -4 & 5 \end{bmatrix}$$

and label them as B and C.

(C)   Now **Author** $A + B$ and **Simplify** gives

$$A + B = \begin{bmatrix} 1 & -2 & 4 \\ -2 & 3 & 7 \end{bmatrix}.$$

Alternatively try **Author** #1 + #3 and **Simplify**. This shows the matrices to be added and their sum, see figure 11.2.

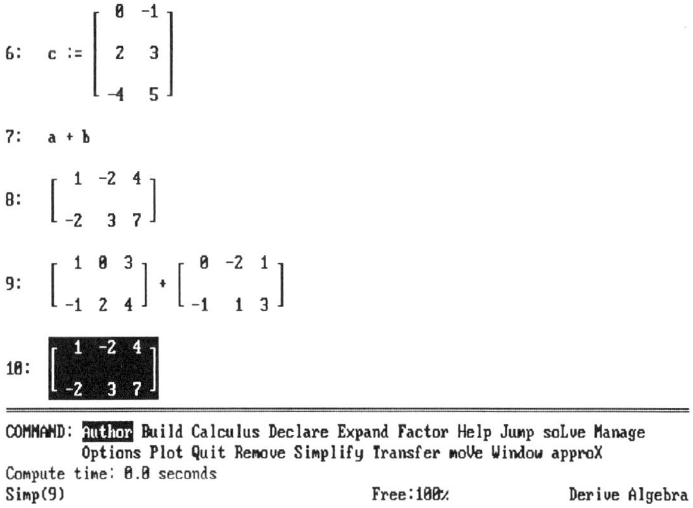

6:    c := $\begin{bmatrix} 0 & -1 \\ 2 & 3 \\ -4 & 5 \end{bmatrix}$

7:    a + b

8:    $\begin{bmatrix} 1 & -2 & 4 \\ -2 & 3 & 7 \end{bmatrix}$

9:    $\begin{bmatrix} 1 & 0 & 3 \\ -1 & 2 & 4 \end{bmatrix} + \begin{bmatrix} 0 & -2 & 1 \\ -1 & 1 & 3 \end{bmatrix}$

10:    $\begin{bmatrix} 1 & -2 & 4 \\ -2 & 3 & 7 \end{bmatrix}$

COMMAND: **Author** Build Calculus Declare Expand Factor Help Jump soLve Manage
          Options Plot Quit Remove Simplify Transfer moVe Window approX
Compute time: 0.0 seconds
Simp(9)   Free:100%   Derive Algebra

Figure 11.2

(D)   To multiply two matrices together you use a 'dot' between the matrices.
       Author $A.C$ and **Simplify** gives

$$AC = \begin{bmatrix} -12 & 14 \\ -12 & 27 \end{bmatrix}.$$

Alternatively try **Author #1.#5** and **Simplify**. This shows the matrices to be
multiplied and their product, see figure 11.3.

10: $\begin{bmatrix} 1 & -2 & 4 \\ -2 & 3 & 7 \end{bmatrix}$

11: a · c

12: $\begin{bmatrix} -12 & 14 \\ -12 & 27 \end{bmatrix}$

13: $\begin{bmatrix} 1 & 0 & 3 \\ -1 & 2 & 4 \end{bmatrix} \cdot \begin{bmatrix} 0 & -1 \\ 2 & 3 \\ -4 & 5 \end{bmatrix}$

14: $\begin{bmatrix} -12 & 14 \\ -12 & 27 \end{bmatrix}$

```
COMMAND: Author Build Calculus Declare Expand Factor Help Jump soLve Manage
         Options Plot Quit Remove Simplify Transfer moVe Window approX
Compute time: 0.0 seconds
Simp(13)                              Free:100%              Derive Algebra
```

Figure 11.3

(E)   What happens if you try $A.B$? Why will DERIVE not simplify the product?

(F)   Use DERIVE to check all your answers to problems 1 and 3 of Exercise 11B.

11.4 SQUARE MATRICES AND THEIR INVERSES

In section 11.3 we said that division of matrices is not defined but an operation similar to division is possible if we use the *inverse* of a matrix.

Suppose that A and B are two $n \times n$ square matrices and that I is the associated unit matrix. If

$$AB = I$$

then the matrix B is called the *inverse matrix* of A and is written as A^{-1} so that

$$AA^{-1} = I \quad \text{and} \quad A^{-1}A = I.$$

Example 11C

Given that $A = \begin{bmatrix} a & b \\ c & d \end{bmatrix}$ and $B = \dfrac{1}{(ad-bc)}\begin{bmatrix} d & -b \\ -c & a \end{bmatrix}$ where $ad - bc \neq 0$ show that

$$AB = BA = I.$$

Solution

$$AB = \frac{1}{(ad-bc)}\begin{bmatrix} a & b \\ c & d \end{bmatrix} \cdot \begin{bmatrix} d & -b \\ -c & a \end{bmatrix}$$

$$= \frac{1}{(ad-bc)}\begin{bmatrix} ad-bc & 0 \\ 0 & -bc+ad \end{bmatrix}$$

$$= \begin{bmatrix} 1 & 0 \\ 0 & 1 \end{bmatrix} = I$$

$$BA = \frac{1}{(ad-bc)}\begin{bmatrix} d & -b \\ -c & a \end{bmatrix}\begin{bmatrix} a & b \\ c & d \end{bmatrix}$$

$$= \frac{1}{(ad-bc)}\begin{bmatrix} ad-bc & 0 \\ 0 & -cb+ad \end{bmatrix}$$

$$= \begin{bmatrix} 1 & 0 \\ 0 & 1 \end{bmatrix} = I \ .$$

This example provides a simple formula for finding the inverse of a 2×2 matrix. However, it also shows that the inverse of a matrix may not always exist. It can be seen that if $ad - bc$ is zero then the matrix B has no meaning. The quantity $(ad–bc)$ associated with the matrix A is called the *determinant* of A. Every square matrix has associated with it a number called its determinant. We will not develop the theory of determinants in this text but one important consequence of the determinant of a matrix is associated with the inverse matrix.

If the determinant associated with a matrix A is zero then the matrix is said to be *singular* and its inverse A^{-1} does not exist.

Example 11D

Show that the matrix $A = \begin{bmatrix} 1 & 2 \\ 2 & 4 \end{bmatrix}$ does not have an inverse.

Solution

The determinant associated with this matrix is $(1 \times 4 - 2 \times 2) = 0$ so that

$$B = \frac{1}{(ad-bc)} \begin{bmatrix} d & -b \\ -c & a \end{bmatrix} \text{ has no meaning.}$$

One use of the inverse matrix is in the solution of simultaneous linear equations. Suppose that we are given a set of n simultaneous equations in the matrix form

$$Ax = b$$

where x and b are $n \times 1$ column vectors and A is an $n \times n$ matrix.

If we multiply both sides of the equation on the left by A^{-1} we obtain

$$A^{-1}Ax = A^{-1}b$$
$$Ix = A^{-1}b$$
$$x = A^{-1}b$$

Hence if we can find A^{-1} we can solve for x.

Example 11E

Solve the linear equations

$$2x + y = 1$$
$$x + 4y = 11$$

Solution

In matrix form the problem can be written as

$$\begin{pmatrix} 2 & 1 \\ 1 & 4 \end{pmatrix}\begin{pmatrix} x \\ y \end{pmatrix} = \begin{pmatrix} 1 \\ 11 \end{pmatrix}$$

so that

$$\begin{pmatrix} x \\ y \end{pmatrix} = \begin{pmatrix} 2 & 1 \\ 1 & 4 \end{pmatrix}^{-1}\begin{pmatrix} 1 \\ 11 \end{pmatrix}.$$

Using the formula of Example 11C gives

$$\begin{pmatrix} 2 & 1 \\ 1 & 4 \end{pmatrix}^{-1} = \frac{1}{7}\begin{pmatrix} 4 & -1 \\ -1 & 2 \end{pmatrix}$$

and then

$$\begin{pmatrix} x \\ y \end{pmatrix} = \frac{1}{7}\begin{pmatrix} 4 & -1 \\ -1 & 2 \end{pmatrix}\begin{pmatrix} 1 \\ 11 \end{pmatrix} = \frac{1}{7}\begin{pmatrix} -7 \\ 21 \end{pmatrix} = \begin{pmatrix} -1 \\ 3 \end{pmatrix}.$$

Hence $x = -1$ and $y = 3$.

Exercise 11C

1.   If possible find the inverse of each of the following matrices:

(a) $A = \begin{pmatrix} 1 & 5 \\ 3 & -2 \end{pmatrix}$ (b) $B = \begin{pmatrix} 0 & 2 \\ -1 & 4 \end{pmatrix}$

(c) $C = \begin{pmatrix} -1 & 2 \\ 6 & -12 \end{pmatrix}$ (d) $D = \begin{pmatrix} 0.3 & -0.4 \\ 1 & 0.7 \end{pmatrix}$

2.   Use the inverse of a matrix to solve the following equations.

(a) $x + 5y = 2$ (b) $0.3x - 0.4y = 1.2$
       $3x - 2y = 1$ $x + 0.7y = 0.8$

(c) $2x + 5y = 19$ (d) $ax + by = r$
       $3x + y = 9$ $cx + dy = s$

3.   Given that $A = \begin{bmatrix} 1 & -1 & 2 \\ 2 & 0 & 1 \\ -1 & 3 & 4 \end{bmatrix}$ and $B = \dfrac{1}{18}\begin{pmatrix} -3 & 10 & -1 \\ -9 & 6 & 3 \\ 6 & -2 & 2 \end{pmatrix}$ show that $B = A^{-1}$ and

hence solve the linear equations

$$\begin{aligned} x - \quad y + 2z &= 2 \\ 2x \quad + \quad z &= -1 \\ -x + 3y + 4z &= 3 \end{aligned}$$

4.   Given that $A = \begin{bmatrix} 0 & 1 & -1 & 2 \\ -1 & 3 & 1 & 1 \\ 4 & -2 & 0 & -1 \\ 1 & 1 & -1 & 2 \end{bmatrix}$ and $B = \begin{pmatrix} -1 & 0 & 0 & 1 \\ -6 & -0.5 & -1.5 & 5.5 \\ 9 & 1.5 & 2.5 & -8.5 \\ 8 & 1 & 2 & -7 \end{pmatrix}$ show

that $A = B^{-1}$ and hence solve the linear equations

$$\begin{aligned} -x_1 \qquad\qquad + \quad x_4 &= 2 \\ -6x_1 - 0.5x_2 - 1.5x_3 + 5.5x_4 &= 0 \\ 9x_1 + 1.5x_2 + 2.5x_3 - 8.5x_4 &= 0.1 \\ 8x_1 + \quad x_2 + 2x_3 - 7x_4 &= 0 \end{aligned}$$

5.   DERIVE can be used to find matrix inverses and determinants.

(A)   Enter the matrix $\begin{pmatrix} 1 & -1 & 2 \\ 2 & 0 & 1 \\ -1 & 3 & 4 \end{pmatrix}$.

Author F3 brings the matrix to the message line. Now type $\wedge(-1)$ and

the DERIVE window shows $\begin{pmatrix} 1 & -1 & 2 \\ 2 & 0 & 1 \\ -1 & 3 & 4 \end{pmatrix}^{-1}$.

Simplify returns the inverse of the original matrix.

(B)   Use DERIVE to find the inverse of each of the following matrices.

(i) $\begin{pmatrix} 2 & 1 \\ -1 & 4 \end{pmatrix}$ (ii) $\begin{pmatrix} 1 & 2 & -1 \\ -1 & 1 & 2 \\ 2 & -1 & 1 \end{pmatrix}$

(iii) $\begin{pmatrix} 1 & 1 & 2 & 3 \\ 1 & 2 & 4 & -1 \\ 2 & 4 & -1 & 1 \\ 0 & 1 & 2 & -1 \end{pmatrix}$ (iv) $\begin{pmatrix} 1 & 2 & 3 \\ 4 & 5 & 6 \\ 7 & 8 & 9 \end{pmatrix}$

(C)   DERIVE can be used to find the determinant associated with a matrix.

Enter the matrix $\begin{pmatrix} 1 & -1 & 2 \\ 2 & 0 & 1 \\ -1 & 3 & 4 \end{pmatrix}$ and label it as a.

Now **Author** DET(a) and **Simplify**. DERIVE gives the answer 18.
Use DERIVE to find the determinants associated with the matrices in
(B).
What can you say about the inverse of the matrix in (iv)?

(D)   Given the set of linear equations

$$2x_1 + x_2 + 3x_3 = 5$$
$$2x_2 + x_3 = 4$$
$$3x_1 + x_2 + 6x_3 = 10$$

(i)   Write the equations in matrix form $Ax = b$;

(ii) Find the inverse of A;

(iii) Hence solve the equations for x_1, x_2 and x_3.

(E)   Use the steps in part (D) to solve the following:

(i) $x_1 + 2x_2 - x_3 = 1$
       $-x_1 + x_2 + 2x_3 = -5$
       $2x_1 - x_2 + x_3 = 4$

(ii) $x + 2y + 3z + w = 4$
       $2x + y + z + w = 3$
       $x + 3y + z + w = 2$
       $y + z + 2w = 1$

(iii) $x_1 + 2x_2 + 3x_3 + x_4 = 5$
       $2x_1 + x_2 + x_3 + x_4 = 3$
       $x_1 + 2x_2 + x_3 = 4$
       $x_2 + x_3 + 2x_4 = 0$

(iv) $-x_1 + x_4 = 2$
       $-6x_1 - 0.5x_2 - 1.5x_3 + 5.5x_4 = 0$
       $9x_1 + 1.5x_2 + 2.5x_3 - 8.5x_4 = 0.1$
       $8x_1 + x_2 + 2x_3 - 7x_4 = 0$

6.   An alternative direct method of solving a set of n simultaneous linear equations using DERIVE is to enter the equations as an $n \times 1$ column vector using the **Declare vectoR** commands.

Consider equations

$$x + 2y = 3$$
$$-x - 3y = 4$$

Declare vectoR

respond to DECLARE VECTOR: Dimension: _ with 2 because you have 2 equations.
respond to VECTOR element: _ with $x + 2y = 3$ and $-x - 3y = 4$.

soLve gives the solution for x and y as a 1×2 row vector. Figure 11.4 shows the DERIVE screen for this activity.

```
1:   [x + 2 y = 3, -x - 3 y = 4]

2:   [x = 17, y = -7]
═══════════════════════════════════════════════════════════
COMMAND: Author Build Calculus Declare Expand Factor Help Jump soLve Manage
         Options Plot Quit Remove Simplify Transfer moVe Window approX
Compute time: 0.1 seconds
Solve(1)                              Free:100%              Derive Algebra
```

Figure 11.4

Use the **Declare vectoR** commands to solve the sets of linear equations in problem 5 parts (D) and (E).

Answers to Exercises

Exercise 1A

2.   (a) $y = 3x$ (b) $y = \frac{4}{3}x - \frac{1}{3}$ (c) $y = x + 4$

    (d) $y = -2x + 6$ (e) $y = \frac{1}{3}x - 1$ (f) $y = -\frac{1}{5}x + 5$

    (g) $y = -3x + 7$ (h) $y = -\frac{4}{3}x + \frac{14}{3}$

Exercise 1B

1.   (a) No   (b) $S = 10R$ (c) No   (d) $t = 1.5v$

2. $k = \frac{5}{8}$, 6.25N

3. $k = \frac{1}{5}$, 500 miles

4.   1.5 cm

5.

Normal Reaction	$R(N)$	19.8	16.2	15.3
Friction	$F(N)$	6.1	4.99	4.7

$F = 0.308N$

6. $P = k\rho T$

Exercise 1C

1.   (a) $y = 4.03846x + 0.676923$ (b) $s = 3.1r - 4.2$

    (c) $v = -9.8t + 20$ (d) $p = 1.72x + 0.309989$

2.   (b) $R = 0.172761T + 44.6183$ (c) $R = 48.07$ ohms

3. $v = 3.26742t - 0.048$
    acceleration $= 3.27$ ms^{-2} (to 3 sf)

Exercise 1D

1.   (a) $x = 5$ (b) $x = 5\frac{1}{2}$ (c) $x = \frac{6}{7}$ (d) $x = \frac{2}{5}$

    (e) $x = 1$ (f) $x = -3$ (g) $x = 2$ (h) $x = \frac{6}{5}$

    (i) $t = 4$ (j) $t = 26$ (k) $x = 1.575$ (l) $x = 1.39273$

2.   45, 47, 49

3.   (i) length = 20 cm, width = 10 cm
    (ii) length = 19 cm, width = 11 cm

4.   (a) $a = 2$ (b) $u = 4$ (c) $u = -2$

Exercise 1E

1.   (a) $x = 10, y = 6$ (b) $x = 3, y = 2$ (c) $x = 6, y = 2$

     (d) $x = \dfrac{38}{17}, y = -\dfrac{20}{17}$ (e) $a = 4, b = -2$ (f) $u = -2, t = 5$

Exercise 1F

2.   (a) $(x+4)(x+5)$ (b) $(x-4)(x-3)$ (c) $(x+1)(x+3)$ (d) $(x+2)(x+3)$

     (e) $(x-1)(x+1)$ (f) $(2x+1)(x-5)$ (g) $(2x+1)(x+2)$ (h) $(5x+1)(x-7)$

     (i) $(3x+2)(x-4)$ (j) $(x-2)(x+2)$

3.   (a) $x = -3$ (repeated)   (b) $x = -2, x = -8$
    (c) $x = 2, x = -2\frac{1}{2}$ (d) $x = 4, x = -4$

4.   (a) $x = 1, x = -7$ (b) $x = -2.52, x = 1.19$ (c) No real solutions

     (d) No real solutions   (e) $x = 0.232, x = 1.434$ (f) No real solutions

     (g) $x = 1, x = -\dfrac{94}{13}$ (h) $x = 2.65, x = -2.65$

Exercise 1H

1.   (a) 0   (b) 1   (c) 3   (d) 4   (e) 5   (f) –7

2.   (a) $\dfrac{x^2}{4}$ (b) $\dfrac{x-6}{2}$ (c) $\dfrac{x^2}{2}$

     (d) $\dfrac{x}{2} - 6$ (e) $\left(\dfrac{x-6}{2}\right)^2$ (f) $\dfrac{x^2 - 6}{2}$

Exercise 1I

1. (a) $f^{-1}(x) = \dfrac{x+10}{6}$ (b) $f^{-1}(x) = \dfrac{x}{4} - 2$ (c) $f^{-1}(x) = \dfrac{x^2 - 7}{2}$

   (d) $f^{-1}(x) = \sqrt[3]{\dfrac{x+5}{8}}$ (e) $f^{-1}(x) = \dfrac{x^2 + 4}{2}$ (f) $f^{-1}(x) = \left(\dfrac{x+5}{4}\right)^2 - 1$

   (g) $f^{-1}(x) = 2((x+5)^2 - 1)$ (h) $f^{-1}(x) = \sqrt[3]{(5x)^2 - 4}$

2. (a) $f^{-1}(x) = \dfrac{2(x+2)}{1-x}$ (b) $f^{-1}(x) = \dfrac{x}{1-x}$ (c) $f^{-1}(x) = \sqrt[3]{\dfrac{2x}{1-x}}$

   (d) $f^{-1}(x) = \dfrac{3}{2x}$ (e) $f^{-1}(x) = \dfrac{2(x+3)}{1-x}$ (f) $f^{-1}(x) = \sqrt[3]{\dfrac{3-x}{2x-1}}$

3. (a) 0 (b) $\frac{1}{2}$ (c) $4\frac{1}{2}$ (d) $-5\frac{7}{8}$ (e) $-6\frac{1}{2}$ (f) $\frac{5}{2}$

Exercise 1J

(a) $x = 1, y = 0$ (b) $x = -\frac{2}{3}, y = 0$ (c) $x = 1, x = 2, y = 0$

(d) $x = 0, x = -2, y = 0$ (e) $x = -1, y = 2$ (f) $x = 4, y = 1$

(g) $x = -2, x = 4, y = 0$ (h) $x = 5, x = -4, y = 1$

Exercise 2A

1. (a) x^5 (b) a^{11} (c) $128a^5$

    (d) $3x^3$ (e) c^4 (f) $3^8 = 6561$

    (g) a^6 (h) $a^8 b^{12}$ (i) $\dfrac{1}{5}$

    (j) $\dfrac{1}{2^4} = \dfrac{1}{16}$ (k) a^2 (l) $a^{-1} = \dfrac{1}{a}$

    (m) mg^3 (n) $a^{3x} - a^{-x}$ (o) $\dfrac{a^6}{x^4 y^2}$

2. (a) $x^{\frac{1}{2}}$ (b) $x^{\frac{5}{2}}$ (c) $x^{\frac{1}{2}}$
    (d) x^6 (e) x^2 (f) $x^{-\frac{1}{2}}$

3. (a) 3   (b) 2   (c) 2
    (d) 1000   (e) 10   (f) $\frac{1}{2}$
    (g) 1   (h) 5   (i) 1

4. (a) 320000   (b) 0.0001473   (c) 9810
    (d) 0.00000103   (e) 192300000000   (f) 3168000
    (g) 59050000000   (h) 4740000

5. (a) 3.75×10^8 joules   (b) 3.16×10^{-11} m
    (c) 2.7×10^{-3} m   (d) 1.82×10^{-22} joules
    (e) 6.3×10^5 N   (f) 3.15×10^7 s

Exercise 2B

1. (a) 1   (b) 7.389   (c) 0.04979
    (d) 4.953   (e) 1.234   (f) 1.822
    (g) 2.718   (h) 1.649   (i) 54.60
    (j) 0.8187

3. n $\left(1+\dfrac{1}{n}\right)^n$

    1   2
    1.5   2.1516574
    2   2.25
    3   2.37037
    4   2.44140
    5   2.48832
    10   2.59374
    100   2.70481
    1000   2.71692
    10000   2.71816

4.   (a) 1.196051   (b) 8.16616
     (c) 1210.28   (d) −0.0216608

Exercise 2C

1.   (a) 0.30103   (b) 1   (c) 0.6234
     (d) −0.1549   (e) 0.6931   (f) 1.435
     (g) −0.35667   (h) 0   (i) 3
     (j) 2   (k) 0   (l) 1

3.   (a) $x = 0.491$ (b) $x = -0.699$ (c) $x = 0.531$

     (d) $x = 1.629$ (e) $t = 2.416$ (f) $t = -0.755$

     (g) $t = 5.474$ (h) $x = 66.69$ (i) $x = \dfrac{10^6}{3}$

     (j) $t = 4.440$

4. Q_0 is the maximum change
     $t = 0.014$ seconds
     As $t \to \infty$, $Q \to Q_0$

5. $t = 2.77259$

6.   5599 years
     18599 years
     24198 years

7.   (a)   86071, 74082, 47237, 22313
       (c)   4.62 km
       (d)   15.35 km

Exercise 2D

1.   (a) $3\ln a + 2\ln b$ (b) $2\ln x + \ln 0.1$
       (c) $\ln 3.7 + \ln t - \ln a$ (d) $\ln 3 + \ln p + 2\ln(x+y)$
       (e) $\frac{1}{2}\log 2 + \log s - \log t$ (f) $\log 0.2 + 2\log r$
       (g) $1 + \log x + 2\log y$ (h) $\ln\frac{1}{2} + \ln a + 2\ln t$
       (i) $\log 3 + \log(a+b) - \log s$

2.   (a) $\log 8$ (b) $\log(20x)$

       (c) $\ln\left(\dfrac{10ab}{c}\right)$ (d) $\ln(a^3 b^5)$

       (e) $\ln\left(\dfrac{x^2}{y^4}\right)$ (f) $\log\left(\dfrac{a^{1.5}x}{y}\right)$

3.   (a) $x = 135.2$ (b) $x = 2.008$ (c) $x = 2.686$

Exercise 2E

1.   (a)   linear law 6
       (b)   power law 1
       (c)   exponential law 2,3

2.   (a) $y = 3.43x^{1.9}$ (b) $t = 1.5s^{-2}$

3.   (a) $v = 4e^{-0.2u}$ (b) $y = 6.8e^{1.2x}$

4. $T = 0.2R^{1.5}$

5.   S.V.P. $= 0.61e^{0.067T}$

6. $I = 0.00076e^{24.5V}$

7. $p = 100e^{-0.15h}$

8. $k = 5.1 \times 10^{12}e^{-30300/T}$

Exercise 3A

1. (a) $\dfrac{13\pi}{36} = 1.13\,\text{radians}$ (b) $\dfrac{\pi}{10} = 0.314\,\text{radians}$

    (c) 85.9° (d) 28.6°

    (e) 160.4° (f) $\dfrac{5\pi}{6} = 2.618\,\text{radians}$

    (g) 286.5° (h) $\dfrac{31\pi}{20} = 4.869\,\text{radians}$

    (i) 57.296° (j) $\dfrac{\pi}{180} = 0.0175\,\text{radians}$

2.

Degrees	0	30	45	60	90	120	135	150	180
Radians	0	$\dfrac{\pi}{6}$	$\dfrac{\pi}{4}$	$\dfrac{\pi}{3}$	$\dfrac{\pi}{2}$	$\dfrac{2\pi}{3}$	$\dfrac{3\pi}{4}$	$\dfrac{5\pi}{6}$	π

Degrees	210	225	240	270	300	315	330	360
Radians	$\dfrac{7\pi}{6}$	$\dfrac{5\pi}{4}$	$\dfrac{4\pi}{3}$	$\dfrac{3\pi}{2}$	$\dfrac{5\pi}{3}$	$\dfrac{7\pi}{4}$	$\dfrac{11\pi}{6}$	2π

3. (a) 0.4 radians (b) 1.75 radians
4. $\frac{2}{3} = 0.67\,\text{radians}$; 48 cm^2
5. 1.2 cm

DERIVE ACTIVITY 3A

(E) (a) 10.30 m (b) 29.1°

(F) 6.71 miles, 26.6°

(G) 58.4 m

(H) (a) 36.9° (b) 63.4° (c) 63.5°

DERIVE ACTIVITY 3B

(C) $a = 5.76$ cm, $b = 9.22$ cm   (D) $A = 39.1°$, $C = 64.1°$, area $= 32.1$ cm^2

(E)   (a)   622 m   (b)   65 m

(F)   (a) $PR = 125.7$ m, $QR = 178.2$ m (b)   area $= 9284$ m^2

(G)   170° 34.9 cm   (H)   4.27 nautical miles, bearing 333°

Exercise 3B

2.   (a)   odd   (b)   even   (c)   odd   (d)   neither
     (e)   neither   (f)   even   (g)   even   (h)   odd
     (i)   even   (j)   even   (k)   even

Exercise 3C

2.   (a)   57.289962   (b)   572.95721   (c)   5729.5779
     (d)   −5729.5779   (e)   −572.95721   (f)   −57.289962

3.   At every odd multiple of 90° i.e. ±90°, ±270°, ±450°, ±630° .....

Exercise 3D

1.   (a)   (i)   3   (ii)   5   (iii)   0   (iv)   0
     (b)   (i)   4   (ii)   1   (iii) $\dfrac{\pi}{2}$ (iv)   0
     (c)   (i)   1   (ii)   3   (iii) $-\dfrac{2\pi}{3}$ (iv)   0
     (d)   (i)   1   (ii)   3   (iii)   0   (iv)   2
     (e)   (i)   1   (ii)   3   (iii) $-\frac{2}{3}$ (iv)   0
     (f)   (i)   2   (ii)   4   (iii) $\dfrac{\pi}{4}$ (iv)   1
     (g)   (i)   1   (ii) $\frac{1}{2}$ (iii) 2π (iv)   −3
     (h)   (i)   0.1   (ii) 2π (iii) $\frac{3}{2}$ (iv)   0.5

3.   (a) $6\cos 2t$ (b) $3\sin(4t+\pi)$ (c) $\tan 5t + 3$

      (d) $10\sin\left(5t - \dfrac{\pi}{2}\right)$ (e) $0.5\cos(3t-2)-1$

      (f) $\dfrac{\pi}{10}$ (g) $\dfrac{10}{\pi}$ (h) 200π

5.   (a)   (i) $\dfrac{2\pi}{3}$ (ii) $\dfrac{3}{2\pi}$

      (b)   (i) $\dfrac{2\pi}{5}$ (ii) $\dfrac{5}{2\pi}$

      (c)   (i) $\dfrac{1}{4}$ (ii)   4

      (d)   (i)   4   (ii) $\dfrac{1}{4}$

      (e)   (i) $\dfrac{\pi}{4}$ (ii) $\dfrac{4}{\pi}$

Exercise 3E

1. $f(t) = 80\sin 120\pi t$

2.   6 cm, 1.5 Hz, at the central position, 1.85 cm to the left of the central position

3. $T = 11\sin\left(\dfrac{\pi t}{6} + \dfrac{\pi}{2}\right) + 16$, in June $T = 21.5,°C$, in January $T = 6.47°C$

4. $a = 6.5$, $w = \dfrac{2\pi}{12.4} = 0.5067$, $\alpha = \dfrac{\pi}{2}$
      h = 5.97 m, 4.48 m, 2.26 m

Exercise 3F

1.   23.6°   2.   66.4°   3.   21.8°
4.   −53.1°   5.   143.1°   6.   −38.7°

Exercise 3G

1. (a) $x = -331.965°, -208.035°, 28.034°, 151.965°$
   (b) $x = -260.212°, -99.788°, 99.788°, 260.212°$
   (c) $x = -305.54°, -125.54°, 54.46°, 234.46°$
   (d) $x = -312.84°, -47.16°, 47.16°, 312.84°$
   (e) $x = -243.43°, -63.43°, 116.57°, 296.57°$
   (f) $x = -117.13°, -62.87°, 242.87°, 297.13°$

3. (a) $\theta = -71.94°, 71.94°, 288.06°, 431.94°$
   (b) $\theta = -41.67°, 138.33°, 318.33°, 498.33°$
   (c) $\theta = 64.16°, 115.84°, 424.16°, 475.84°$

Exercise 3H

1. (a) 51.7 °   (b) −46.4 °   (c) −61.6 °
   (d) 81.4 °   (e) 3.44 °   (f) 161.8 °

2. (a) −1.29   (b) −0.100   (c) 1.51
   (d) 1.91   (e) 0.464   (f) 0.110

3. (a) 0.995   (b) 0.995   (c) 0.75
   (d) −0.314   (e) 0.298   (f) −0.848
   (g) 0.8   (h) −0.55   (i) 0.75

Exercise 3I

1. $t = -5.20168, -1.0815, 1.0815, 5.20168$
2. $t = -6.11946, -3.30532, 0.16373, 2.97786$
3. $t = -3.23135, -0.0897581, 3.05183, 6.19343$
4. $t = -3.00113, -0.140461, 3.28205, 6.13858$
5. $t = -5.36809, -2.22649, 0.91510, 4.05669$
6. $t = -4.72238, -1.56079, 1.56079, 4.72238$

Exercise 3J

3. (a) $x = 45°, 135°, 215°, 315°$
    (b) $\theta = 15.3°, 164.7°$
    (c) $x = 35.9°, 144.1°, 226.9°, 313.1°$
    (d) $x = 60°, 131.8°, 228.2°, 300°$
    (e) $x = 53.6°, 147.5°, 212.5°, 306.4°$
    (f) $x = 61°, 119°, 270°$
    (g) no solutions
    (h) $\theta = 39.2°, 140.8°, 219.2°, 320.8°$

Exercise 3K

1. $\cos(A+B) = 0$
    $\sin(A+B) = 1$
    $\tan(A+B)$ is undefined
    $A + B = 90°$

2. (a) $\dfrac{7}{25}$ (b) $\dfrac{15}{17}$ (c) $\dfrac{7}{24}$ (d) $\dfrac{15}{8}$

    (e) $-\dfrac{304}{425}$ (f) $\dfrac{304}{297}$ (g) $\dfrac{87}{425}$ (h) $\dfrac{416}{87}$

    (i) $\dfrac{527}{625}$ (j) $\dfrac{240}{289}$

3. (a) 0.866   (b) 0.6   (c) 1.73   (d) 0.75
    (e) 0.393   (f) −0.427   (g) −0.1196   (h) −8.30
    (i) −0.5   (j) 0.96

4. (b) (i) $-\sin A$ (ii) $-\cos A$ (iii) $-\sin A$

    (iv) $-\cos A$ (v) $\sin A$ (vi) $\dfrac{\tan A - 1}{1 + \tan A}$

5. (a) $-\sin 7x$ (b) $2\sin 40°\cos 20°$ (c) $-2\sin 40°\sin 10°$
    (d) $\frac{1}{2}(\sin 80° + \sin 20°)$ (e) $-\frac{1}{2}(\cos 50° - \cos 30°)$ (f) $\sin 29°$
    (g) $2\cos 40°\sin 5°$ (h) $\sin(45° + x)$

6. (b) $4\cos^3 A - 3\cos A$ (c) $4\sin A \cos A - 8\sin A \cos^3 A$

Exercise 3L

1. (a) $13\sin(x+22.6°)$ (b) $13\sin(x-22.6°)$

    (c) $\sqrt{5}\sin(x+63.4°)$ (d) $10\sin(x-53.1°)$

    (e) $\sqrt{26}\sin(x-11.3°)$ (f) $\sqrt{45}\sin(x-26.6°)$

2. $A\sin x + B\cos x = R\cos(x-\alpha)$ where $R = \sqrt{A^2 + B^2}$ and $\alpha = \tan^{-1}\left(\dfrac{A}{B}\right)$

3. (a) $13\cos(-22.6°)$ (b) $\sqrt{45}\cos(x-116.6°)$

    (c) $\sqrt{5}\cos(x-116.6°)$ (d) $\sqrt{10}\cos(x-18.4°)$

4. (a) $x = 4.9°, 129.9°$ (b) $x = -157.4°, 22.6°$

    (c) $x = -36.8°, 90°$ (d) $x = 143.1°$

    (e) $x = -180°, 22.6°, 180°$ (f) $x = 74.8°, 158.4°$

Exercise 4A

1. (a) $u_n + 3n + 1$ (b) $u_n = n^2 + 2$
    (c) $u_n = n^2 - 1$ (d) $u_n = 5n - 1$

2. (a) $2, 4, 8, 16, 32$ no limit
    (b) $1, \dfrac{1}{3}, \dfrac{1}{9}, \dfrac{1}{27}, \dfrac{1}{81}$ limit $= 0$
    (c) $-2, 4, -8, 16, -32$ no limit
    (d) $-\dfrac{1}{2}, \dfrac{1}{4}, -\dfrac{1}{8}, \dfrac{1}{16}, -\dfrac{1}{32}$ limit $= 0$
    (e) $3, 2\dfrac{1}{2}, 2\dfrac{1}{3}, 2\dfrac{1}{4}, 2\dfrac{1}{5}$ limit $= 2$
    (f) $1.1, 2.01, 3.001, 4.0001, 5.00001$ no limit

3. (a) $u_n = 3^{n-1}$ (b) $u_n = 19 - 5n$
    (c) $u_n = 5 + (0.1)^n$ (d) $u_n = \dfrac{n}{n+2}$
        limit $= 5$ limit $= 1$

Exercise 4B

1. (a) $4 + 9 + 16 + 25 = 54$ (b) $1 + \dfrac{1}{2} + \dfrac{1}{3} + \dfrac{1}{4} + \dfrac{1}{5} + \dfrac{1}{6} = \dfrac{49}{20} = 2.45$
    (c) $2 + 6 + 10 + 14 = 32$ (d) $2 + 4 + 6 + 8 + 10 = 30$
    (e) $1 + \dfrac{1}{4} + \dfrac{1}{9} = \dfrac{49}{36}$ (f) $-2 + 5 + 24 + 61 = 88$

2. (a) $\displaystyle\sum_{i=1}^{5} (21 - 5n)$ (b) $\displaystyle\sum_{i=1}^{4} 2(3^{i-1})$
    (c) $\displaystyle\sum_{i=1}^{6} \dfrac{1}{2} i^2$ (d) $\displaystyle\sum_{i=1}^{5} (-2)^{i-1}$

3. only (e) to $\dfrac{\pi^2}{6}$

Exercise 4C

1.   (a) 64   (b) -3 (c) 19
            $u_n = 9n - 8$ $u_n = 5 - n$ $u_n = 2n + 3$

2.   (a) $d = 3, n = 28$ (b) $d = 4, n = 18$ (c) $d = -7, n = 16$

3.   (a) 852   (b) -78 (c) 1515

4.   (a) 100   (b) 205

5.   420

6.   4

7. $a = 6, d = 4$

Exercise 4D

1.   (a) (i) 1.2   (ii) 20.6391   (iii) 65.9963
      (b) (i) 0.8   (ii) 0.174483   (iii) 5.40948
      (c) (i) -0.5 (ii) 0.0033203   (iii) -1.1289
      (d) (i) 1.6   (ii) 13744   (iii) 13983.3

2.   (a) $66\frac{2}{3}$ (b) does not exist   (c) 10   (d) $-\dfrac{10}{19}$

3. $2\frac{1}{3}$ m

4.   2 or $\dfrac{1}{3}$

Exercise 4E

1.   (a) $x^4 + 4x^3 + 6x^2 + 4x + 1$ (b) $8 + 12x + 6x^2 + x^3$

    (c) $1 - 5x + 10x^2 - 10x^3 + 5x^4 - x^5$ (d) $625 + 1500x + 1350x^2 + 540x^3 + 81x^4$

    (e) $125 - 150x + 60x^2 - 8x^3$ (f) $x^5 - 5x^4y + 10x^3y^2 - 10x^2y^3 + 5xy^4 - y^5$

    (g) $u^4 + 4u^2 + 6 + \dfrac{4}{u^2} + \dfrac{1}{u^4}$ (h) $16x^4 + 32x^2 + 24 + \dfrac{8}{x^2} + \dfrac{1}{x^4}$

2. (a) 1920000   (b) –96   (c) –343   (d) –12

3. (a) 1716   (b) –42240   (c) 9773.16   (d) –6.08402 × 10^{-6}
    (e) $6435a^8b^7$ (f) $-3432a^7b^7$

4. (a) $(1+x)^6$ (b) $(0.5+x)^4$ (c) $(1-0.5x)^4$

Exercise 4F

1. (a) $1 - 2x + 3x^2 - 4x^3$ (b) $1 - 3x + 6x^2 - 10x^3$

    (c) $1+\dfrac{3}{2}x+\dfrac{3}{8}x^2 -\dfrac{1}{16}x^3$ (d) $1-\dfrac{x}{2}+\dfrac{3x^2}{8}-\dfrac{5}{16}x^3$

2. (a) $|x|<\dfrac{1}{2}$ (b) $|x|<3$ (c) $|x|<\dfrac{1}{4}$ (d) $|x|<\dfrac{1}{5}$

3. (a) $1+\dfrac{3}{2}x-\dfrac{9}{8}x^2 +\dfrac{27}{16}x^3$ $|x|<\dfrac{1}{3}$

    (b) $1 + 4x + 12x^2 + 32x^3$ $|x|<\dfrac{1}{2}$

    (c) $1 - 4x + 16x^2 - 64x^3$ $|x|<\dfrac{1}{4}$

    (d) $1-x+\dfrac{3}{2}x^2 -\dfrac{5}{2}x^3$ $|x|<\dfrac{1}{2}$

4. 0.989949, 1.0955

5. (b) $1-\dfrac{6}{x}+\dfrac{27}{x^2}-\dfrac{108}{x^3}$ (c) $|x|>3$ (d) $\dfrac{1}{x^2}-\dfrac{6}{x^3}+\dfrac{27}{x^4}-\dfrac{108}{x^5}$

6. (a) $\dfrac{1}{2}-\dfrac{x}{4}+\dfrac{x^2}{8}-\dfrac{x^3}{16}$ (b) $\dfrac{1}{4}-\dfrac{x}{8}+\dfrac{x^2}{16}-\dfrac{x^3}{32}$

    (c) $\dfrac{1}{25}-\dfrac{2}{125}x^2 +\dfrac{3}{625}x^2 -\dfrac{4}{3125}x^3$ (d) $\sqrt{2}\left(1+\dfrac{1}{4}x-\dfrac{1}{32}x^2 +\dfrac{1}{128}x^3\right)$

Exercise 5A

1. ±1.50   2. −2.31   3. 1.935   4. −0.567

5. (a) 2.15443   (b) 2.71442

Exercise 5B

1. 6.1926, no

2. 0.8074, no

3. (b) 3.1004, 3.107
   (c) 2nd converges faster, 3.1073

4. (a) 2.8284, 3.1623
   (c) $2x^2 = a$, 2.2361

5. (a) 0
   (b) $x_{n+1} = \sqrt{\sin x_n}$, $x_{n+1} = \sin^{-1}(x_n^2)$, 1st converges
   (c) 0.877

6. (a) 1.74
   (b) 0.739
   (c) −1.841, 1.146

Exercise 6A

1. (a) $7x^6$ (b) $\dfrac{1}{3}x^{-\frac{2}{3}}$ (c) $-3x^{-4}$ (d) $-\dfrac{5}{2}x^{-\frac{7}{2}}$

(e) $12x^2$ (f) $9x^{\frac{1}{2}}$ (g) $-\dfrac{0.7}{x^2}$ (h) $9x^8 + 4x^3 + 1$

(i) 0 (j) $30x^4 - 2$ (k) $-3x^{-2} - x^{-\frac{3}{2}}$ (l) $15 - 12x^3 + 14x^6$

(m) $24x^2 + 2x - 3$ (n) $6.3x^8 + 0.6x^{-3}$ (o) $-0.3x^{-4} + 5.7x^2$

(p) 0

2. (a) 12 (b) 5 (c) 1

(d) -4 (e) $-\dfrac{1}{9}$ (f) $\dfrac{1}{4}$

Exercise 6B

1. (a) $6, y = 6x - 9$ (b) $12, y = 12x + 16$

(c) $\dfrac{5}{16}, y = \dfrac{5}{16}x - \dfrac{1}{8}$ (d) $\dfrac{1}{4}, y = \dfrac{1}{4}x + 1$

(e) $\dfrac{5}{4}, y = \dfrac{5}{4}x - \dfrac{1}{4}$ (f) $0.461827, y = 0.461827x + 0.791704$

(g) $-\dfrac{100}{91}, y = -\dfrac{100}{9}x + \dfrac{20}{3}$ (h) $-1.11976, y = -1.11976x + 2.05094$

(i) $-96, y = -96x - 144$ (j) $135, y = 135x - 729$

(k) $-2, y = -2x + 1$ (l) $-7, y = 10 - 7x$

2. (a) -1, decreasing; 15, increasing
   (b) 12, increasing; 0, stationary; 3 increasing
   (c) -7, decreasing; -3.25, decreasing
   (d) -1.21837, decreasing; 1.11679, increasing
       the curve reaches a local minimum

Exercise 6C

1. 40mph
    Getting to and from the M1
    Road works on the M1
    Stops at service stations

2. 1.15ms^{-2}

3. (a) 35ms^{-1} (b) 25ms^{-1} (c) 17.5ms^{-1}

    (b) 40ms^{-1}, 30ms^{-1}, 20ms^{-1}, 10ms^{-1}, 0ms^{-1}
         The stone stops moving instantaneously at its highest point

    (c) -10ms^{-2} for all times; the acceleration is directed downwards

4. 1.41 fish per second; 0.67 fish per second

5. $\dfrac{dA}{dr} = 2\pi r$

6. (a)

x	$f(x)$	$\Delta f(x)$	av. rt
1.7	7.87	1.12	5.6
1.6	7.28	0.53	5.3
1.55	7.0075	0.2575	5.15
1.51	6.80030	0.0502999	5.02999
1.501	6.75500	0.05	5
1.5001	6.7505	0.0005	5

    Instantaneous rate of change of f when $x = 1.5$ is 5

    (b) $\dfrac{df}{dx} = 5$

7.

r	$p(r)$	$\Delta p(r)$	av. rt
2.2	4.78181	0.78181	3.90905
2.1	4.39523	0.39523	3.95229
2.01	4.03995	0.03995	3.995
2.001	4.00399	0.00399	3.99
2.0001	4.0004	0.0004	4

The instantaneous rate of change of p when $r = 2$ is 4

Exercise 6D

1. (a) $2x$ (b) 1   (c) $6x - 4$ (d) $10x + 2$

    (e) $3x^2$ (f) $6x^2 + 3$ (g) $-\dfrac{1}{x^2}$ (h) 0

2. $g'(t) = \lim\limits_{h \to 0} \dfrac{g(t+h) - g(t)}{h}$

    (a) 1   (b) $6t - 4$ (c) $15t^2 - 6$ (d) 0

Exercise 6E

1. (a) $(0, -12)$ minimum
    (b) $(0, 9)$ maximum
    (c) $(-1.25, -6.125)$ minimum
    (d) $(-0.167, 2.083)$ maximum
    (e) $(0, 3)$ point of inflexion
    (f) $(-0.0972, 5.049)$ maximum $(3.431, -16.9)$ minimum)
    (g) $(-0.76929, 10.8765)$ maximum
    (h) $(-1, -4)$ maximum $(1, 0)$ minimum

2. (e) $(0, 3)$
    (f) $(1.67, -5.926)$
    (g) $(0.1835, 2.7542)$ $(1.8165, -10.3097)$

Exercise 6F

1. (a) $6x$ (b) $-12x + 100x^3 - 30x^4$ (c) $2 + \dfrac{1}{4}x^{-\frac{3}{2}} + \dfrac{5}{16}x^{-\frac{9}{4}}$

    (d) $\dfrac{2}{x^3}$ (e) $-\dfrac{1}{2}x^{-\frac{3}{2}} + 6x$ (f) $8 - 210x^4$

    (g) 8   (h) 0   (i) $-\dfrac{1}{4}t^{-\frac{3}{2}} + \dfrac{3}{8}t^{-\frac{7}{4}} + \dfrac{4}{3}t^{-\frac{7}{3}}$ (j) $n(n-1)x^{n-2}$

3. (a) $20x^3 - 24x^2;\ 60x^2 - 48x;\ 120x - 48$

    (b) $-12x + 100x^3 - 30x^4;\ -12 + 300x^2 - 120x^3;\ 600x - 360x^2$

    (c) $\dfrac{2}{x^3}; -\dfrac{6}{x^4}; \dfrac{24}{x^5}$ (d) $-\dfrac{1}{4}t^{-\frac{3}{2}} - 12t;\ -12 + \dfrac{3}{8}t^{-\frac{5}{2}}; -\dfrac{15}{16}t^{-\frac{7}{2}}$

Exercise 6G

(a)   (0, –11) minimum
(b)   (1, 7) inflexion
(c)   (0.382, –1) min; (1.5, 0.5625) max; (2.618, –1) min
(d)   (1.42264, 0.385) max; (2.57735, –0.385) min
(e)   (1, 2) max, (–1, –2) min
(f)   none

Exercise 6H

1.   1m from one end

2.   Base radius = 1.3365 m
     Height = 2.673 m
     Area required = 33.7 m^3

3.   If $S = x + y$ (and $xy = k$), $\dfrac{dS}{dx}$ is a minimum when $x = y = \sqrt{k}$
     For $k = 10^4$, $x = y = 10^2$ Kmol^{2}dm^{-6}

4.   (i) $v = \dfrac{1}{3}\sqrt{\dfrac{T}{3a}}$

5.   800 m^2

7.   0.08

8. $r = 20/3$ and $V = \pi\,(20/3)^3$

Exercise 6I

1. (a) $4e^{4x}, 16e^{4x}$ (b) $-7e^{-7x}, 49e^{-7x}$

    (c) $2e^{0.5x}, e^{0.5x}$ (d) $-2.6e^{-1.3x}, 3.38e^{-1.3x}$

    (e) $\dfrac{1}{x}, -\dfrac{1}{x^2}$ (f) $\dfrac{3}{x}, -\dfrac{3}{x^2}$

    (g) $\pi\cos(\pi x), -\pi^2\sin(\pi x)$ (h) $2\cos(2x), -4\sin(2x)$

    (i) $13.02\cos(3.1x), -40.36\sin(3.1x)$ (j) $-4\sin 4x, -16\cos 4x$

    (k) $-0.2\sin(0.2x), -0.04\cos(0.2x)$ (l) $-3\pi\sin(2\pi x), -6\pi^2\cos(2\pi x)$

    (m) $0.03e^{0.1x} - 0.35\cos(0.5x), 0.003e^{0.1x} + 0.175\sin(0.5x)$

    (n) $-12\sin 3x - 12\cos 4x, -36\cos 3x + 48\sin 4x$

    (o) $-0.1e^{-0.1x} + 0.1e^{0.1x}, 0.01e^{-0.1x} + 0.01e^{0.1x}$

    (p) $\dfrac{1}{x} - \dfrac{6}{x} = -\dfrac{5}{x}, \dfrac{5}{x^2}$

2. (a) $y = -5.98x + 3.134$ (b) $y = x - 1$
    (c) $y = 7.389x - 7.389$ (d) $y = -0.1x + 0.4$

4. (a) $-500\ °Cs^{-1}$ (b) $-3.37\ °Cs^{-1}$
5. $\dfrac{dP}{dt} = -9.03e^{-2.1t}$

6. (a) $0.21ms^{-1}; 0.16ms^{-1}$
    (b) $0ms^{-2}; -0.0947ms^{-2}$

Exercise 6J

1. (a) $e^{2x}\left(2\sqrt{x}+\dfrac{1}{2\sqrt{x}}\right)$ (b) $e^{3x}(3x^2+2x)$

(c) $x^4e^{-2x}(5-2x)$ (d) $2x\cos 2x + \sin 2x$

(e) $\dfrac{\cos(\pi x)}{2\sqrt{x}}-\pi\sqrt{x}\sin(\pi x)$ (f) $e^{3x}(15x^3+15x^2)=15x^2e^{3x}(x+1)$

(g) $\sec^2 x$ (h) $1+\ln x$

(i) $\dfrac{2x\cos 2x-2\sin 2x}{x^3}$ (j) $\dfrac{e^{-3x}(1-6x)}{2\sqrt{x}}$

(k) $\dfrac{-3\sin 2x\sin 3x-2\cos 3x\cos 2x}{\sin^2 2x}$ (l) $\dfrac{2e^{2x}(x-1)-2e^{-2x}(x+1)}{x^3}$

(m) $15(3x-1)^4$ (n) $2(4x+1)^{-\frac{1}{2}}$

(o) $-\pi\sin(\pi x-3)$ (p) $2xe^{x^2}$

(q) $\dfrac{2x}{x^2+1}$ (r) $-2\tan 2x$

(s) 0 (t) $e^{-2x}(3x^2\cos 3x+2x(1-x)\sin 3x)$

(u) $12\sin(1-4x)$ (v) $\dfrac{3(6x^2-1)^2(6x^2+1)}{x^4}$

(w) $e^{5x}(0.7\cos 0.7x+5\sin 0.7x)$ (x) $\dfrac{2}{x}$

(y) $\sec x\tan x$ (z) $-2(4x+1)^{-\frac{3}{2}}$

2. (a) $(0.7937, 1.88988)$ minimum (b) none

(c) $(0,0)$ minimum (d) none

(e) $t=n\pi+0.294$, $n=0,\pm 1,\pm 2,$ ..... maximum
$t=n\pi+1.865$, $n=0,\pm 1,\pm 2,$ ..... minimum

(f) none (g) none (h) inflexions at $n\pi$

3.   (c)   1.15 radians = 65.9°

4.   (a)   9 cm   (b)   1 s   (c)   1 cms^{-1} increasing   (d)   12 cm

5.   (a)   (−2, 3) max; (2, 0.33) min

     (b) $x = 2n\pi - \dfrac{\pi}{4}$ min

                              $y = \pm 2.33$

           $x = 2n\pi + \dfrac{3\pi}{4}$ max

     (c) $x = 2n\pi \pm \pi$ points of inflexion

     (d) $x = n\pi + \dfrac{\pi}{4}$ max when n is even
                             min when n is odd

6.   (a) $X = R$

Exercise 7A

1. Estimates obtained using interval mid-points. True values in brackets.

   (a) 8.9775 (9)
   (b) 3.74625 (4)
   (c) 0.50028 (0.5)
   (d) 3.59660 (3.69328)

2. (a) $\int_0^3 x^2 dx$ (b) $\int_1^2 x^3 dx$

   (c) $\int_0^{\pi/3} \sin x \, dx$ (d) $\int_{-1}^3 e^{-2x} \, dx$

3. (b) distance = 2.37 m (using 10 sub-intervals)

4. (a) 12.92 miles
   (b) measure the speed at shorter time intervals

Exercise 7B

1. (a) $x^4 + c$ (b) $\dfrac{1}{2}x^6 + c$ (c) $\dfrac{2}{3}x^{\frac{3}{2}} + c$

   (d) $\dfrac{13}{3}x^3 - \dfrac{7}{4}x^4 + c$ (e) $\dfrac{1}{2}x^6 + \dfrac{1}{2}x^4 - \dfrac{1}{2}x^2 + 4x + c$

   (f) $6x + \dfrac{3}{2}x^2 - \dfrac{2}{3}x^3 + c$ (g) $\dfrac{16}{3}x^3 + 8x^2 + 4x + c$

   (h) $\dfrac{x^3}{3} - x^2 + x + c$ (i) $-\dfrac{1}{x} + c$ (j) $\dfrac{1}{0.3}x^{0.3} + c$

   (k) $-\dfrac{1.7}{1.3}x^{-1.3} + c$ (l) $3\ln(x) + c$ (m) $-x^{-5} - 3x^{-1} + c$

   (n) $\dfrac{2}{1.3}x^{1.3} + \ln(x) + c$ (o) $\dfrac{1}{2}x^{18} - \dfrac{2}{5}x^5 - 3x^{-1} + c$

2. (a) $\dfrac{1}{2}e^{2x} + c$ (b) $-\dfrac{1}{5}e^{-5x} + c$ (c) $10e^{0.1x} + c$

   (d) $\dfrac{3}{4}e^{4x} + c$ (e) $e^{6x} + c$ (f) $1.8e^{-0.5x} + c$

   (g) $\dfrac{4}{3}e^{3x} + \dfrac{3}{2}e^{-2x} + c$ (h) $\dfrac{0.6}{3.1}e^{3.1x} + 3e^{-0.3x} + c$

3. (a) $-\dfrac{1}{5}\cos(5x)+c$ (b) $\dfrac{2}{3}\sin(1.5x)+c$

(c) $-\dfrac{4}{3}\cos(3x)+c$ (d) $-\dfrac{3}{2}\cos(2x)-\dfrac{2}{3}\sin(3x)+c$

(e) $-\dfrac{2}{\pi}\cos(\pi x)+c$ (f) $\dfrac{0.5}{\pi}\sin(3\pi x)+c$

(g) $-\dfrac{2}{w}\cos(wx)+c$ (h) $\dfrac{1.5}{7}\sin(7x)-0.15\cos(2x)+c$

(i) $10e^{0.1x}-\dfrac{2}{\pi}\cos(\pi x)+c$ (j) $x^3-7e^{-0.6x}+\dfrac{1}{0.9}\sin(0.9x)+c$

(k) $\dfrac{1}{3}\ln(x)-0.1\cos(5x)+c$

4. (a) $\dfrac{7}{3}(2.33)$ (b) 7.5   (c) $\dfrac{26}{3}(8.67)$

(d) $\dfrac{2}{3}e^3-\dfrac{2}{3}(12.7236)$ (e) 17.301   (f) $25(e^{0.2x}-e^{-0.2x})\,(10.0667)$

(g) 2   (h) 1   (i) $\dfrac{178}{3}-70e^{-0.4}\,(12.4109)$

5. (a) $\dfrac{13}{15}(0.8667)$ (b) 2

(c) 0.0664   (d) $2e^2-4\;(10.7781)$

6. (a) $\dfrac{1}{3}x^3-\dfrac{3}{2}x^2+2x+c$ (b) $\dfrac{1}{4}x^4+\dfrac{2}{3}x^3+c$

(c) $-\dfrac{1}{2}x^{-2}-x^{-1}+c$ (d) $\dfrac{1}{3}x^3-x^{-1}+c$

(e) $-ax^{-1}+bx+c$ (f) $\dfrac{1}{3}ax^3+\dfrac{1}{2}bx^2+cx+d$

7. $y=2x+x^2-\dfrac{1}{3}x^3-\dfrac{5}{3}$

8. $y=x+\dfrac{1}{3}x^3+1$

9.   (a) $\dfrac{1}{3}t^3 + c$ (b) $\dfrac{3}{2}t^2 + t + c$ (c) $\dfrac{1}{4}t^4 + \dfrac{2}{3}t^3 + c$

    (d) $-\dfrac{1}{t} + c$ (e) $\dfrac{5}{8}t^8 + t^4 - \ln(t) + c$ (f) $\dfrac{t^3}{3} - t^2 + t + c$

    (g) $\dfrac{1}{3}at^3 + \dfrac{1}{2}bt^2 + ct + d$ (h) $\dfrac{1}{2}e^{2t} + c$ (i) $20e^{0.1t} + c$

    (j) $-\dfrac{2}{\pi}\cos(\pi t) + c$ (k) $\ln(t) + c$ (l) $50\sin(0.1t) + c$

    (m) $-\cos(t) + \sin(t) + c$ (n) $-\dfrac{1}{v} + c$ (o) $2p^{\frac{1}{2}} + c$

    (p) $\dfrac{1}{3}u^3 + \dfrac{3}{2}u^2 + 8u + c$ (q) $\ln(w) + c$ (r) $\dfrac{1}{2}e^{2u} + c$

    (s) $\dfrac{2}{3}y^{\frac{3}{2}} + c$

10. (a) $v = \dfrac{1}{3}t^3 + 2$ (b) $v = \dfrac{1}{2}t^2 + t - \dfrac{1}{2}$ (c) $v = 3\sin(t) + c$

11. (a) $V = mgx + c$

    (b) $V = \dfrac{k}{2}x^2 - klx + c$

Exercise 7C

2.   (a) $\dfrac{16}{3}(5.33)$ (b) $\dfrac{11}{12}(0.9167)$

    (c) 2   (d) 39.33

3.   (a) -4

       area between $y = x - 2$, $y = 0$, $x = -1$ and $x = 1$ is 4

    (b) 1.3863 (2ln(2))

       area between $y = \dfrac{1}{x}$, $y = 0$, $x = 0.5$ and $x = 2$ is 1.3863

(c) $0.6667 \left(\dfrac{2}{3} \right)$

not an area because $y = x^2 - x$ is negative between $x = 0$ and $x = 1$ and positive between $x = 1$ and $x = 2$

(d) $2(e^1 - e^{-1})$ (4.7008)

area between $y = 2e^{-x}$, $y = 0$, $x = -1$ and $x = 1$ is 4.7008

(e)   0

not an area because $y = \sin(x)$ is negative between $x = -\dfrac{\pi}{2}$ and $x = 0$ and

positive between $x = 0$ and $x = \dfrac{\pi}{2}$

(f) $1 - 3e^{-1}$ (−0.1036)

not an area because $y = 3e^{-x} - 2$ is negative between $x = \ln(1.5)$ and $x = 1$

4.   (a)   area = 15.0

$$\int_{-3}^{3} f(x)dx = 15.0$$

(b)   area = 15.0

$$\int_{-3}^{3} f(x)dx = 0$$

(c)   area = 5.4

$$\int_{0}^{1} f(x)dx = 2.7$$

$$\int_{1}^{2} f(x)dx = -2.7$$

$$\int_{0}^{2} f(x)dx = 0$$

Exercise 7D

1.   (a) $\dfrac{(4x^2-1)^2}{16}+c$ (b) $\dfrac{2}{3}(x-1)^{\frac{3}{2}}+c$

    (c) $-\dfrac{1}{2}\cos(2x+1)+c$ (d) $\dfrac{2}{3}\ln(3x+2)+c$

    (e) $\dfrac{1}{10}\ln(5x^2-2)+c$ (f) $\dfrac{1}{2}e^{2x+1}+c$

    (g) $\dfrac{1}{16}(4x+1)^4+c$ (h) $\dfrac{-1}{(1+x)}+c$

    (i) $\dfrac{1}{5}\ln(5x-3)+c$ (j) $\dfrac{1}{2}e^{x^2}+c$

    (k) $\dfrac{1}{1.3}(1+x)^{1.3}+c$ (l) $-\ln(\cos x)+c$

    (m) $\dfrac{2}{3}(x^2+x+3)^{1.5}+c$ (n) $\ln(x^2+3x-5)+c$

    (o) $\dfrac{1}{24}(4t-11)^6+c$ (p) $\dfrac{1}{2(3-2r)}+c$

    (q) $\dfrac{1}{2}\sin(2y+\pi)+c$ (r) $\dfrac{1}{2}\ln(1+t^2)+c$

    (s) $-\dfrac{1}{3}\cos^3(\theta)+c$ (t) $\dfrac{1}{10}\sin^5(2\theta)+c$

2.   (a) $-\dfrac{1}{3}$ (−0.3333)   (b) $\dfrac{1}{5}\ln(6)$ (0.3584)

    (c) $\dfrac{1}{2}\ln\!\left(\dfrac{5}{3}\right)$ (0.2554)   (d) $\left(5\sqrt{5}-1\right)/3$ (3.3934)

    (e) $\dfrac{2}{3}\ln(3)$ (0.7324)   (f) $\dfrac{1}{2}(e^1-1)$ (0.8591)

    (g) $\dfrac{1}{3}(e^1-e^{-8})$ (0.9060)   (h) 34

Exercise 7E

(a) $\dfrac{1}{2}\arcsin(2x) + c$ (b) $\dfrac{1}{2}\arctan(2t) + c$

(c) $\arcsin\left(\dfrac{x}{3}\right) + c$ (d) $\dfrac{1}{3}\arctan(3x) + c$

(e) $\dfrac{\pi}{6}(0.5236)$ (f) $\dfrac{\pi}{4}(0.7854)$

Exercise 7F

1.   (a) $-x\cos x + \sin x + c$ (b) $xe^x - e^x + c$

(c) $\dfrac{1}{2}x^2\,e^{2x} - \dfrac{1}{2}xe^{2x} + \dfrac{1}{4}e^{2x} + c$ (d) $-\dfrac{1}{3}x\cos 3x + \dfrac{1}{9}\sin 3x + c$

(e) $t^2e^t - 2te^t + 2e^t + c$ (f) $\dfrac{1}{2}x^2\ln x - \dfrac{1}{4}x^2 + c$

(g) $x\ln(x) - x + c$

2.   (a)   1   (b) -0.5 (c)   8.6328   (d)   0.1905

(e) $(1+e^{\pi/2})/2$ (2.9052)   (f) $\dfrac{3}{13}(e^{-2\pi/3} + 1)$ (0.2592)

Exercise 7G

(a) $\ln(x-1) - \ln(x+1) + c$ (b) $\dfrac{1}{5}\ln(2x-1) - \dfrac{1}{5}\ln(x+2) + c$

(c) $\dfrac{5}{4}\ln(x-3) - \dfrac{5}{4}\ln(x+1) + c$ (d) $3\ln(t-3) - 3\ln(t-2) + c$

(e) $\dfrac{1}{4}\ln(2v-1) + \dfrac{1}{4}\ln(2v+3) + c$ (f) $5\ln(5) + 3\ln(3) - 16\ln(2)$ (0.2527)

(g) $\dfrac{23}{6}\ln(x+3) - \dfrac{3}{2}\ln(x-1) - \dfrac{1}{3}\ln(x) + c$ (h) $\dfrac{5}{4}\ln(2) - \dfrac{3\pi}{8}(-0.3117)$

(i) $\dfrac{2}{\sqrt{19}}\arctan\!\left((2x-9)/\sqrt{19}\right)$ (j) $\ln(3) - 2\ln(2) + 0.5$ (0.2123)

Exercise 7H

1. (a) $\dfrac{1}{2}\sin^2 x + c$; or $-\dfrac{1}{2}\cos^2 x + c$; or $-\dfrac{1}{4}\cos 2x + c$

(b) $\ln(x^3 - 3) + c$

(c) $\dfrac{1}{2}e^{2x} + \dfrac{1}{4}\cos 4x + c$

(d) 2

(e) $-\dfrac{1}{4}x\sin(2x) + \dfrac{1}{4}\cos 2x + \dfrac{1}{4}x^2 + c$

(f) $\dfrac{1}{2}\arctan\left(\dfrac{1}{2}\right)$ (0.2318)

(g) $\dfrac{1}{4}\ln(3)$ (0.2747)

(h) $\dfrac{1}{2}\ln(4 + t^2) + c$

(i) $\dfrac{1}{8}(e^4 - 1)$ (6.70)

(j) $\ln(t - 2) - \ln(t + 3) + c$

(k) $\dfrac{1}{2}u^2 + 2\ln(u) + c$

(l) $\dfrac{1}{2}\arcsin(v) + \dfrac{1}{2}v\sqrt{1 - v^2} + c$

(m) $\dfrac{1}{5}u^5 + \dfrac{1}{3}u^3 + c$

(n) $\dfrac{1}{4}(e^7 - e^{-1})$ (274.07)

(o) $(2 - u^2)\cos(u) + 2u\sin(u) + c$

(p) $\dfrac{1}{3}te^{3t} - \dfrac{1}{9}e^{3t} + c$

2, 875 metres

3. $T = \dfrac{k}{2(2t + 3)} + T_0$

where k is the constant of proportionality and T_0 is a constant of integration

4. $V = \dfrac{15}{2}x^2 - 20x + c$

5. $y = \ln(x - 2) - \ln(x - 1) + 1 + \ln(2)$

6. total area = 16.5

7.   area $= \dfrac{1}{6}$ (0.167)

8.   26.3856 (using right ends of intervals)

9.   (a)   area $= 0.5$

    (b) $\int_{1}^{2}(2x-3)dx = 0$

10. $t = 0.5919\dfrac{u}{g}$

11. $H = 0.4472$

Exercise 8A

1. (a) $1.89549, 0, -1.89549$ (b) -2.1663
   (c) $1.85722, 4.5364$ (d) 0.7391
   (e) $-1, 1.3532$ (f) 0.5671

2. (a) 2.1544 (b) 1.5157 (c) $-1.5850, 1.5850$

3. The method does not converge for $x = -1$ and $x = 1$

4. $(0.1444, 1.0719)$ Maximum
   $(3.2617, -16.4608)$ Minimum

5. $k = 0.1594$

Exercise 8B

1. (a) $1 + x + \dfrac{x^2}{2}$ (b) $1 - \dfrac{x^2}{2} + \dfrac{x^4}{24}$ (c) $x^2 - \dfrac{x^6}{6} + \dfrac{x^{10}}{120}$

   (d) $1 - x + x^2$ (e) $x + \dfrac{x^3}{3} + \dfrac{2x^5}{15}$ (f) $1 - \dfrac{x}{2} - \dfrac{x^2}{8}$

   (g) $-\dfrac{x^2}{2} - \dfrac{x^4}{12} - \dfrac{x^6}{45}$ (h) $1 + x^2 + \dfrac{x^4}{2}$

Exercise 8C

1. (a) $0.5 - 0.866\left(x - \dfrac{\pi}{3}\right) - 0.25\left(x - \dfrac{\pi}{3}\right)^2$

   (b) $1 + 2\left(x - \dfrac{\pi}{4}\right) + 2\left(x - \dfrac{\pi}{4}\right)^2$ (c) $1 - (x - 1) + (x - 1)^2$

2. (a) $2x - \dfrac{4}{3}x^3 + \dfrac{4}{15}x^5$ (b) $3x - \dfrac{9}{2}x^3 + \dfrac{81}{40}x^5$

3. (b) $-\dfrac{x^2}{2} - \dfrac{x^4}{12} - \dfrac{x^6}{45} - \dfrac{17x^8}{2520} - \dfrac{31x^{10}}{14175}$ (c) -0.05003

4. 1.4408

5. 1.4618

Exercise 8D

1. (a) trapezoidal: 2.9814
        Simpson: 2.9274

    (b) trapezoidal: 1.5656
        Simpson: 1.5708

    (c) trapezoidal: 0.4058
        Simpson: 0.4055

2. (a) 0.746546   (b) 0.746528   (c) 0.746526
    The answer is 0.7465 to four decimal places

3. 0.845 (using 32 subintervals)

4. trapezoidal: 8.81
    Simpson: 8.83

6. Integrals (c) and (d) do not exist

Exercise 9A

1. (a) Let i be the current (milliamps) at time t (seconds)

$$\frac{di}{dt} = -\frac{i}{80}$$

(b) Let T be the temperature (°C) at time t (minutes)

$$\frac{dT}{dt} = -\frac{4}{45}(T-15)$$

(c) Let h be the depth (metres) at time t (hours)

$$\frac{dh}{dt} = -\frac{1}{35}(1+2h)$$

(d) Let m be the mass (grams) deposited at time t (seconds)

$$\frac{dm}{dt} = km(20-m)$$

2. It takes π seconds to change from 20 cm to 15 cm

3. Assume that the rate of change of temperature is proportional to the difference between the temperature of the can of drink (T) and the surroundings.

$$\frac{dT}{dt} = k(22-T)$$

4. $\dfrac{dh}{dt} = \dfrac{4.8}{\pi h^2}$

Exercise 9B

1. (a) $y = Ae^x$ (b) $y = Ae^{\frac{1}{2}x^2}$ (c) $y = Ae^{\frac{1}{4}x^4}$

2. (a) $m = Ae^{-5t}$ (b) $m = 10e^{-5t}$

3. (a) $y = \dfrac{10}{1-10\ln(x)}$ (b) $y = \dfrac{6}{3-2x^3}$ (c) $y = \sqrt{100 + \tfrac{2}{3}x^3}$

4. 13.9 minutes

5. (a) $y = \ln\left(e^{10} + \dfrac{1}{2} - \dfrac{1}{2}\cos(2x)\right)$ (b) $y = -\ln\left(e^{-1} - \dfrac{1}{3}x^3\right)$

    (c) $y = \sqrt{2e^x + 2}$

6. (a) $y = \dfrac{2\left(A\,e^{4x^3/3} + 1\right)}{1 - A\,e^{4x^3/3}}$ (b) $y = A(x-2) - 2$

    (c) $y = \dfrac{A\,e^{x^3/3}}{1 - A\,e^{x^3/3}}$

7. $v = \dfrac{1}{1.025e^{0.2t} - 1}$

8. (a) and (b)

Exercise 9C

1. Euler's method: $\theta(6) \approx 114.954$
    Percentage error: 7.7%

2. Euler's method: $y(3.7) \approx 86.8358$
    Exact solution: $y(3.7) = 82.0900$
    Percentage error: 5.8%

3. Euler's method: $y(0.3) \approx 1.0304$
    Exact solution: $y(0.3) = 1.04712$
    Percentage error: 1.6%

4. 8.85957

Exercise 10A

1.

Complex Number	Real Part	Imaginary Part	Complex Conjugate
$2 - 3i$	2	-3	$2 + 3i$
$6 + 2i$	6	2	$6 - 2i$
$1.73 - 2.19i$	1.73	-2.19	$1.73 + 2.19i$
$1.7 + 4.6i$	1.7	4.6	$1.7 - 4.6i$
$-5.17 + 1.03i$	-5.17	1.03	$-5.17 - 1.03i$
$-4i$	0	-4	$4i$
$17i$	0	17	$-17i$
$x - yi$	x	$-y$	$x + yi$
$-p + qi$	$-p$	q	$-p - qi$

2. $x = -1 + \sqrt{2}i, \ x = -1 - \sqrt{2}i$

$x = 3i, \ x = -3i$

$x = \dfrac{1}{4} + \dfrac{\sqrt{7}}{4}i, \ x = \dfrac{1}{4} - \dfrac{\sqrt{7}}{4}i$

$x = -\dfrac{1}{3} + \dfrac{\sqrt{5}}{3}i, \ x = -\dfrac{1}{3} - \dfrac{\sqrt{5}}{3}i$

4. (a) $x^2 - 4x + 13 = 0$
    (b) $100x^2 - 40x + 29 = 0$
    (c) $x^2 - 6x + 34 = 0$
    (d) $x^2 - 2ax + (a^2 + b^2) = 0$

Exercise 10B

1. (a) modulus $= 1$, argument $= \dfrac{\pi}{2}$

    polar form $= \cos\dfrac{\pi}{2} + i\sin\dfrac{\pi}{2}$

    (b) modulus $= \sqrt{2}$, argument $= \dfrac{\pi}{4}$

    polar form $= \sqrt{2}\left(\cos\dfrac{\pi}{4} + i\sin\dfrac{\pi}{4}\right)$

(c)   modulus = $\sqrt{13}$, argument = -0.983

   polar form = $\sqrt{13}\left(\cos(-0.983) + i\sin(-0.983)\right)$

   $\qquad\qquad = \sqrt{13}\left(\cos 0.983 - i\sin 0.983\right)$

(d)   modulus = $\sqrt{17}$, argument = -1.816

   polar form = $\sqrt{17}\left(\cos(-1.816) + i\sin(-1.816)\right)$

   $\qquad\qquad = \sqrt{17}\left(\cos 1.816 - i\sin 1.816\right)$

(e)   modulus = 5, argument = -2.214

   polar form = $5(\cos(-2.214) + i\sin(-2.214))$

   $\qquad\qquad = 5(\cos 2.214 - i\sin(2.214))$

(f)   modulus = 13, argument = 1.966

   polar form = $13(\cos 1.966 + i\sin 1.966)$

(g)   modulus = 1, argument = 0

   polar form = $1(\cos 0 + i\sin 0)$

(h)   modulus = 3, argument = $-\dfrac{\pi}{2}$

   polar form = $3\left(\cos\left(-\dfrac{\pi}{2}\right) + i\sin\left(-\dfrac{\pi}{2}\right)\right)$

   $\qquad\qquad = 3\left(\cos\dfrac{\pi}{2} - i\sin\dfrac{\pi}{2}\right)$

(i)   modulus = 4, argument = π

   polar form = $4(\cos(-\pi) + i\sin(-\pi))$

   $\qquad\qquad = 4(\cos\pi - i\sin\pi)$

(j)   modulus = 2, argument = -1.047

   polar form = $2(\cos(-1.047) + i\sin(-1.047))$

   $\qquad\qquad = 2(\cos 1.047 - i\sin 1.047)$

(k)   modulus = 2, argument = -2.094

   polar form = $2(\cos(-2.094) + i\sin(-2.094))$

   $\qquad\qquad = 2(\cos 2.094 - i\sin 2.094)$

(l)   modulus = 6.35, argument = –1.36
       polar form = 6.35(cos(–1.36) + $i$sin(–1.36))
       = 6.35(cos1.36 – $i$sin1.36)

2.   (a) $x = -1 + i, x = -1 - i$

$$x = \sqrt{2}\left(\cos\frac{3\pi}{4} + i\sin\frac{3\pi}{4}\right), x = \sqrt{2}\left(\cos\frac{3\pi}{4} - i\sin\frac{3\pi}{4}\right)$$

      (b) $x = \sqrt{8}i, x = -\sqrt{8}i$

$$x = \sqrt{8}\left(\cos\frac{\pi}{2} + i\sin\frac{\pi}{2}\right), x = \sqrt{8}\left(\cos\frac{\pi}{2} - i\sin\frac{\pi}{2}\right)$$

3.   (a)   5   (b)   –0.927   (c)   5.099
      (d)   9.19   (e)   –1.96   (f)   111.803

Exercise 10C

1.   (a) $3+5i, -3-3i, -4+3i, \dfrac{4}{25}+\dfrac{3}{25}i, -1, -6-5i$

      (b) $-2+3i, -4+5i, 1+7i, -\dfrac{7}{2}+\dfrac{1}{2}i, -7-24i, -11+14i$

      (c) $2, 2i, 2, i, 2i, 1 + 5i$

      (d) $-5-3i, -7+i, -8+11i, -\dfrac{4}{5}-\dfrac{13}{5}i, 35+12i, -20+i$

      (e) $8+i, -2+3i, 17+7i, \dfrac{1}{2}+\dfrac{1}{2}i, 5+12i, -1+8i$

      (f) $2+8i, -8-16i, 33-56i, -\dfrac{63}{169}+\dfrac{16}{169}i, 7+24i, -19-36i$

2.   (a) $2 - 3i$ (b) $3 + 2i$ (c) $\dfrac{3}{13}+\dfrac{2i}{13}$ (d) $\dfrac{3}{13}-\dfrac{2}{13}i$

4.   (a) $a = 3\cos\theta, b = -3\sin\theta$

      (b) $a = \dfrac{2\cos\theta + \sin\theta + 2}{2(\cos\theta + 1)}, b = \dfrac{-(\cos\theta - 2\sin\theta + 1)}{2(\cos\theta + 1)}$

5. $\text{Re}(z) = \dfrac{1}{5}$, $\text{Im}(z) = \dfrac{7}{5}$

6.   (b) $\left|z^n\right| = \left|z\right|^n$, $\arg(z^n) = n\arg(z)$

      $z^n = r^n(\cos(n\theta) + i\sin(n\theta))$

Exercise 10D

1.   (a) $5(\cos 0.927 + i\sin 0.927) = 5e^{0.927i}$

      (b) $\sqrt{2}\left(\cos\left(\dfrac{\pi}{4}\right) - i\sin\left(\dfrac{\pi}{4}\right)\right) = \sqrt{2}e^{-\frac{\pi}{4}i}$

      (c) $\cos\left(\dfrac{5\pi}{6}\right) - i\sin\left(\dfrac{5\pi}{6}\right) = e^{-\frac{5\pi}{6}i}$

      (d) $13(\cos 1.176 - i\sin 1.176) = 13e^{-1.176i}$

      (e) $2\sqrt{2}\left(\cos\dfrac{3\pi}{4} + i\sin\dfrac{3\pi}{4}\right) = 2\sqrt{2}e^{\frac{3\pi}{4}i}$

      (f) $0.762(\cos 1.976 - i\sin 1.976) = 0.762e^{-1.976i}$

      (g) $\dfrac{1}{5}(\cos 0.93 - i\sin 0.93)$, $\dfrac{1}{5}e^{-0.93i}$

      (h) $\dfrac{1}{\sqrt{5}}(\cos 1.11 - i\sin 1.11)$, $\dfrac{1}{\sqrt{5}}e^{-1.11i}$

      (i) $0.109(\cos 0.391 + i\sin 0.391)$, $0.109e^{0.391i}$

2.   (a) -1 (b) $3i$ (c) -1 (d) $1.53 + 1.29i$
      (e) $3.980 - 0.399i$ (f) $0.636 + 0.63i$

3.   (a)   modulus = 4, argument = $\dfrac{\pi}{2}$ (b)   modulus = 27, argument = $\dfrac{\pi}{2}$

      (c)   modulus = 108, argument = π (d)   modulus = $\dfrac{4}{27}$, argument = 0

4.   (a)   modulus = 16, argument = 1.4   (b)   modulus = 0.125, argument = -0.3

      (c)   modulus = 2, argument = 1.1   (d)   modulus = 128, argument = 1.7

5. $5(\cos 0.927 - i\sin 0.927)$, $1526(\cos 0.719 - i\sin 0.719)$

6. $\sqrt{2}\left(\cos\dfrac{\pi}{4} + i\sin\dfrac{\pi}{4}\right)$, $64(\cos 3\pi + i\sin 3\pi) = -64$

7. $\cos 3\theta = \cos^3\theta - 3\sin^2\theta\,\cos\theta = 4\cos^3\theta - 3\cos\theta$
      $\sin 3\theta = 3\sin\theta\,\cos^2\theta - \sin^3\theta = 3\sin\theta - 4\sin^3\theta$

Exercise 10E

		Square Roots	**Cube Roots**
1.	(a)	$1.09868 - 0.455089i$	$1.08421 - 0.290514i$
		$-1.09868 + 0.455089i$	$-0.290514 + 1.08421i$
			$-0.793700 - 0.793700i$
	(b)	$1.45534 + 0.343560i$	$1.29207 + 0.201294i$
		$-1.45534 - 0.343560i$	$-0.820363 + 1.01832i$
			$-0.471711 - 1.21961i$
	(c)	$1.22474 + 1.22474i$	$1.24902 + 0.721124i$
		$-1.22474 - 1.22474i$	$-1.24902 + 0.721124i$
			$0 - 1.44224i$
	(d)	$2 + 0i$	1.58740
		$-2 + 0i$	$-0.793700 + 1.37472i$
			$-0.793700 - 1.37472i$
	(e)	$0.707106 + 1.22474i$	$0.965155 + 1.09112i$
		$-0.707106 - 1.22474i$	$-1.18393 - 1.09112i$
			$0.218782 + 1.09112i$
	(f)	$2 - 3i$	$1.86444 - 1.43269i$
		$-2 + 3i$	$0.308530 + 2.33100i$
			$-2.17297 - 0.898307i$

2. (a) $2^{\frac{1}{5}}\left(\cos\left(\dfrac{-\pi/6+2k\pi}{5}\right)+i\sin\left(\dfrac{-\pi/6+2k\pi}{6}\right)\right)$ $k=0,1,2,3,4$

(b) $2^{\frac{2}{3}}\left(\cos\left(\dfrac{\pi/4+2k\pi}{3}\right)+i\sin\left(\dfrac{\pi/4+2k\pi}{3}\right)\right)^{4}$

$=2^{\frac{2}{3}}\left(\cos\left(\dfrac{\pi+8k\pi}{3}\right)+i\sin\left(\dfrac{\pi+8k\pi}{3}\right)\right)$ $k=0,1,2$

(c) $5^{-\frac{2}{3}}\left(\cos\left(\dfrac{2.214+2k\pi}{3}\right)+i\sin\left(\dfrac{2.214+2k\pi}{3}\right)\right)^{-2}$

$=5^{-\frac{2}{3}}\left(\cos\left(\dfrac{4.428+4k\pi}{3}\right)-i\sin\left(\dfrac{4.428+4k\pi}{3}\right)\right)$ $k=0,1,2$

(d) $\cos\left(\dfrac{(2k+1)\pi}{6}\right)+i\sin\left(\dfrac{(2k+1)}{6}\right)$ $k=0,1,2,3,4,5$

3. $-0.471711+1.21961i$
    $-0.820363-1.01832i$

4. (a) $1.21589+0.503639i$
    $-0.503639+1.21589i$
    $-1.21589-0.503639i$
    $0.503639-1.21589i$

(b) $1.11819-0.097829i$
    $0.643817+0.919467i$
    $-0.474372+1.01729i$
    $-1.11819+0.97829i$
    $-0.643817-0.919467i$
    $0.474372-1.01729i$

Exercise 11A

(a) $\begin{bmatrix} 1 & -1 \\ 2 & 3 \end{bmatrix} \begin{bmatrix} x \\ y \end{bmatrix} = \begin{bmatrix} 4 \\ 1 \end{bmatrix}$

(b) $\begin{bmatrix} -2 & 1 \\ 1 & -1 \end{bmatrix} \begin{bmatrix} x \\ y \end{bmatrix} = \begin{bmatrix} -6 \\ 4 \end{bmatrix}$

(c) $\begin{bmatrix} 3 & -6 & 1 \\ -2 & 1 & -3 \\ 1 & 1 & 1 \end{bmatrix} \begin{bmatrix} x \\ y \\ z \end{bmatrix} = \begin{bmatrix} 7 \\ 2 \\ 0 \end{bmatrix}$

(d) $\begin{bmatrix} 4 & -1 & 1 \\ 3 & 1 & -2 \\ -1 & -4 & 1 \end{bmatrix} \begin{bmatrix} x_1 \\ x_2 \\ x_3 \end{bmatrix} = \begin{bmatrix} 1 \\ -3 \\ 5 \end{bmatrix}$

(e) $\begin{bmatrix} 1 & -3 & 1 & 4 & -1 \\ -1 & 0 & -4.1 & 0 & 2 \\ 0.3 & -0.7 & 0 & 4.1 & -1 \\ 1.4 & -1 & 0 & 3.1 & 0 \\ 3 & 1 & -1 & 2 & -1 \end{bmatrix} \begin{bmatrix} x_1 \\ x_2 \\ x_3 \\ x_4 \\ x_5 \end{bmatrix} = \begin{bmatrix} 0 \\ 1.3 \\ 2.7 \\ 0.4 \\ -3.5 \end{bmatrix}$

Exercise 11B

1. (a) $\begin{bmatrix} 1 & 5 & 5 \\ -1 & -2 & 3 \\ -1 & 2 & 2 \end{bmatrix}$

(b) Impossible

(c) $\begin{bmatrix} 0 & -3 & -9 \\ 3 & 0 & -6 \\ 0 & -3 & -6 \end{bmatrix}$

(d) $\begin{bmatrix} 1 & 1 \\ 1 & 1 \end{bmatrix}$

(e) $\begin{bmatrix} -1 & 1 \\ 1 & -1 \end{bmatrix}$

(f) $\begin{bmatrix} 2 & 5 & -5 \\ 3 & -4 & -4 \\ -2 & -1 & -6 \end{bmatrix}$

(g) Impossible

(h) $\begin{bmatrix} -1 & 2 & 1 \\ 1 & 3 & 0 \end{bmatrix}$

(i) Impossible

(j) Impossible

(k) $\begin{bmatrix} -2 & -7 \\ -2 & -7 \\ -1 & -4 \end{bmatrix}$

(l) $\begin{bmatrix} 11 & 2 & 0 & 6 \\ 8 & -1 & -1 & 1 \\ 7 & 2 & 1 & 4 \end{bmatrix}$

(m) Impossible

(n) $\begin{bmatrix} 1 & -2 & 5 \\ -2 & -7 & 0 \end{bmatrix}$

(o) $\begin{bmatrix} 0 & 1 \\ 1 & 0 \end{bmatrix}$

(p) $\begin{bmatrix} 1 & 3 & 0 \\ -1 & 2 & 1 \end{bmatrix}$

(q) $\begin{bmatrix} 1 & 3 & 0 \\ -1 & 2 & 1 \end{bmatrix}$

(r) $\begin{bmatrix} 0 & 1 \\ 1 & 2 \\ -1 & -3 \end{bmatrix}$

2. $Bx = \begin{bmatrix} x_1 + 3x_2 \\ -x_1 + 2x_2 + x_3 \end{bmatrix}$ $Cx = \begin{bmatrix} x_1 + 4x_2 + 2x_3 \\ -2x_2 + x_3 \\ -x_1 + x_2 \end{bmatrix}$ $Ex = \begin{bmatrix} x_2 + 3x_3 \\ -x_1 + 2x_3 \\ x_2 + 2x_3 \end{bmatrix}$

3.   (a) $\begin{bmatrix} -1 & -3 & 1 \\ -1 & 2 & 1 \\ 1 & 0 & 2 \end{bmatrix}$ (b) $\begin{bmatrix} -3 & -7 & 9 \\ -5 & 6 & 7 \\ 3 & 2 & 10 \end{bmatrix}$ (c) $\begin{bmatrix} \frac{1}{2} & 3 & 4 \\ -1 & -1 & \frac{5}{2} \\ -\frac{1}{2} & \frac{3}{2} & 2 \end{bmatrix}$

4. $a = 0, b = 0$

Exercise 11C

1.   (a) $\begin{bmatrix} \frac{2}{17} & \frac{5}{17} \\ \frac{3}{17} & -\frac{1}{17} \end{bmatrix}$ (b) $\begin{bmatrix} 2 & -1 \\ \frac{1}{2} & 0 \end{bmatrix}$ (c)   No inverse

(d) $\begin{bmatrix} \frac{70}{61} & \frac{40}{61} \\ -\frac{100}{61} & \frac{30}{61} \end{bmatrix}$

2.   (a) $x = \dfrac{9}{17}, \quad y = \dfrac{5}{17}$ (b) $x = \dfrac{116}{61}, \quad y = -\dfrac{96}{61}$

     (c) $x = 2,\, y = 3$ (d) $x = \dfrac{dr - bs}{ad - bc}, \quad y = \dfrac{as - cr}{ad - bc}$

3. $x = \dfrac{19}{18}, \quad y = -\dfrac{5}{6}, \quad z = \dfrac{10}{9}$

4. $x_1 = -0.1,\, x_2 = -1.9,\, x_3 = 8,\, x_4 = 1.9$

Index

DERIVE commands

Other books in the "Learning through Computer Algebra" Series

Published by Chartwell-Bratt. Series editor: John Berry, University of Plymouth

Learning Numerical Analysis through DERIVE T. Etchells, J. Berry

This book covers the major numerical methods, and their analysis, for first courses at college and undergraduate level. The relative merits of each method are covered both analytically, providing a thorough grounding in the algebraic approach, and practically, through the tried and tested computer lab-based activities.

DERIVE provides a platform on which to quickly and accurately perform many complicated numerical calculations. Also, DERIVE's ability to algebraically manipulate expressions and perform calculus operations, enhances the investigation of the convergence of numerical methods.

Each chapter includes the development and algebraic analysis of the methods, lab-based activities, ideas for coursework, case studies, exercises and solutions. Free supporting utility files are downloadable via Chartwell-Bratt's web server.

Chapter 1 introduces the basic tool of numerical methods, which is recurrence relations, their solution and ill-conditioning problems. In chapter 2 we use recurrence relations methods that are used in solving equations. Chapter 3 deals with the approximation of functions by polynomials, and in particular the Taylor Polynomial, which is then used extensively in chapter 4 to analyse the errors associated with numerical methods. Chapters 5 and 6 deal with numerical approaches to the calculus of differentiation and integration. In chapter 7 we introduce and analyse numerical methods of solving differential equations. ISBN 0-86238-468-0, 239 pages, 1997.

Learning Linear Algebra through DERIVE, B. Denton

Using DERIVE, this book reinforces theoretical knowledge while making applications more realistic. It can also be used with packages such as Matlab, Maple, Mathematica and Macsyma, requiring only a few adjustments. The book covers a two-semester course, and will be essential reading for undergraduate students of linear algebra. Contents: Introduction to Matrices. Vectors with Applications to Geometry. Systems of Linear Equations. Vector spaces. Linear Transformation. Eigenvectors and eigenvalues. Conclusions. Solutions. ISBN 0-86238-466-4, 296 pages, 1995

Learning Differential Equations through DERIVE by B. Lowe, J. Berry

Available Spring 1997.

Learning Modelling with DERIVE, S. Townend, D. Pountney.

Develops undergraduate mathematical modelling skills with the support of DERIVE. The book provides students with the opportunity to develop both their problem-solving and IT skills through the fully integrated use of DERIVE. The authors' experience of teaching modelling at Liverpool John Moores University shows that the use of DERIVE enhances the students' modelling expertise.

- Gently guides the reader through the problem formulating, solution and revision stages of modelling.
- Provides a wide range assortment of case studies, some of which are developed fully while others have only hints provided. A wide-ranging set of modelling problems is also provided.

Contents: Introduction. Geometric and Trigonometric Models. Algebraic Models. Optimisation-based Models. Calculus-based Models. Discrete Models. Differential Equations-based Models. Statistical & Simulations Models. The Techniques of Dimensional Analysis. ISBN 0-86238-467-2, 256 pages, 1995